CW01506991

Residenti Leaseholders' Handbook

Charles Ward

2006

 Books

A division of Reed Business Information

Estates Gazette
1 Procter Street, London WC1V 6EU

Typeset in Palatino 10/12 by Amy Boyle, Rochester
Printed by Short Run Press Ltd, Exeter

Contents

Table of Cases

Table of Statutes

Table of Regulations

Abbreviations

All ER	All England Law Reports
CLRA	Commonhold and Leasehold Reform Act 2002
EG	Estates Gazette
EGLR	Estates Gazette Law Reports
EGCS	Estates Gazette Case Summaries
LTL	Lawtel
LVT	Leasehold Valuation Tribunal
SDLT	Stamp Duty Land Tax

Introduction

Most residential leaseholders will have paid the full market price for their flats and maisonettes. Yet the law does not recognise them as "owners" in the fullest sense of the word. Someone else will decide how, when, and by whom essential repairs and building maintenance will be carried out — and how much individual leaseholders will be asked to pay for those repairs and maintenance. That same person will decide with whom, for what risks, for what level of cover and for what cost the building is to be insured. That same person will, at their discretion, allow or refuse applications by individual leaseholders to carry out alterations to their homes. In rare cases, that person's consent may also be required before a residential unit can be sold or sublet. If the leaseholder is affected by another leaseholder's failure to comply with their lease, it will be primarily up to the same person to decide what (if any) action should be taken under the lease to bring that other leaseholder into line.

That person is the residential ground landlord.

Many ground landlords are property investors, whose main incentive is to achieve a secure income stream with the minimum of economic risk and administrative effort. In fact until 1987 it was possible for "ground rents" to be traded like stocks and shares above the heads of the leaseholders they affected.

Others are social landlords such as local authorities and housing associations, whose primary aim is to provide housing for people who might not otherwise be able to afford to purchase their own home on the open market. Many of those properties will have been sold or leased to their sitting tenants under right to buy legislation — replacing their tenancy agreement with a long residential lease or freehold.

But the modern trend is for leaseholders to collectively become their own ground-landlords through the setting up of management companies to buy out the landlord's interest and make decisions democratically on behalf of all the participating leaseholders. It is something, which successive governments have encouraged in a series of legislative measures which began with the 1967 Leasehold Reform Act. The process is still ongoing and should, over time, reduce the traditional areas of conflict between landlord and leaseholder. The most recent legislative reforms were introduced by the Commonhold and Leasehold Reform Act 2002 and protect leaseholders against unwarranted threats of forfeiture for unproven lease contraventions or small financial arrears. But such rights are only of value is leaseholders know about them and understand how to use them.

Whatever the identity of the landlord, most leaseholder problems fall into one of the following types:

- poor maintenance and disrepair — this not only reduces the quality of a leaseholders enjoyment of their property but also can directly impact on the value or saleability of the unit itself
- excessive or unfair service charges
- the uncooperative ground-landlord
- anti-social behaviour by other leaseholders. Even something as trivial as inconsiderate parking can spoil the quality of life for other leaseholders.

Thankfully most owners of flats and maisonettes are able to enjoy their properties without such problems. But when they occur they can be serious, complex and expensive to resolve — if they can be resolved at all.

Until recent legislative reforms, there was little leaseholders could do if faced with a difficult or exploitive landlord. The law was not on their side. But it is directly as a result of those same legislative reforms that residential leaseholders now have the right:

1. to be consulted on major capital items of repair and maintenance before the work is carried out or contracted
2. to challenge unfair or excessive service charges
3. to be given first right of refusal if landlords intend to sell their interest in the property to a third person
4. to collectively buy out the landlords interest and become their own landlord

5. to take over the landlord's responsibilities as regards management of the building within which the flats or maisonettes are situated

The downside is that to implement such rights, leaseholders must act collectively and by majority. Unless leaseholders can first agree among themselves and satisfy other statutory criteria, they may be no further forward.

Anyone searching the Internet will quickly discover a mass of free information from various organisations, including the Leasehold Advisory Service. This in itself will provide leaseholders with a useful overview of their rights, including their statutory rights to challenge service charges, extend leases and buy out landlords' interests.

This book is intended to take the process one step further by giving residential leaseholders and their advisers information in sufficient practical detail to enable them to assess their position before, if necessary, going on to take formal professional advice. In simpler cases, the information provided in this book may be all that leaseholders or their advisers require to resolve the issue in hand.

There are of course many learned and respected volumes dealing in depth with these issues. But such books are mainly written by lawyers, to be read by other lawyers.

While it is hoped that many high street solicitors as well as law students will find this book a useful source of reference in their research, it is written primarily for residential leaseholders and their advisers as well as anyone else who may be affected, either personally in a professional capacity, by the problems and issues described in this book.

This is reflected in the non-technical style which the author has tried to use — combined with sufficient legal detail to make it an authoritative source of advice. The author has also endeavoured to take the law out of the "textbook" and put it into a practical context.

Much use is made of real life case scenarios, many taken from landmark court decisions. Where cases are quoted, the case reference will also be given to enable further research, if required. It must also be remembered that behind each decided case there is a human situation from which we can also learn.

Extensive reference is made throughout the book to the Commonhold and Leasehold Reform Act 2002 (abbreviated to the CLRA) which has significantly improved the position of residential leaseholders. Although the CLRA reforms are being introduced piecemeal, it has been assumed for the purposes of this book that the Act is fully in force. To avoid duplication, reference has only been made to English regulations.

Welsh regulations, which cover the same matters, are published separately and are accessible from the government websites. Scottish leasehold law is entirely different and is not covered in this book.

The first part of the book is intended to help readers understand the basic structure of a residential lease and how it works. It also addresses the issues of assignment (sale), underletting and physical alterations — and what legal redress is available to residential leaseholders when a landlord defaults on its obligations. The book then moves on to the contentious issue of service charges and what leaseholders can do if they think they are being treated unfairly. The second half of the book addresses the more complex statutory issues of acquiring the new right to manage, buying out a freehold, extending a lease — and what happens when a lease expires. The penultimate chapter deals with "dispute resolution" and compares the jurisdiction of the county court with those of the leasehold valuation tribunal which — if current trends continue — could one day become the venue for almost all residential leaseholder disputes.

These issues are more complex because there are at least four statutory regimes giving leaseholders the right to buy out a landlord's interest; two separate regimes which empower leaseholders to control the management of their blocks; and several "legal" ways in which disputes can be resolved. Faced with a choice of statutory procedures, there are many situations where leaseholders (and their advisers) will need to make informed decisions. The information — and case examples — given in this book will assist that decision.

Understanding a Residential Lease

What is a lease?

It is important to differentiate between a lease (or tenancy) and other forms of residential occupancy, such as a licence agreement. The distinction is crucial.

The rights set out in this book will only apply where there is a relationship of landlord and tenant. A lodger in someone else's home may be required to vacate at any time on reasonable notice. A guest in a hotel room will have no rights in that room beyond the terms of their booking. The issue of what constitutes a "tenancy" came before the House of Lords in the 1985 case of *Street* v *Mountford* [1985] 1 EGLR 128 when the Law Lords were called upon to decide whether an occupancy arrangement, described in the documentation as a "personal licence", was in reality a tenancy protected under the Rent Act 1977. The Law Lords ruled that it was.

They said that in the absence of the special circumstances, a tenancy arose whenever there was a grant of *exclusive* possession for a fixed or periodic term at a stated rent. In other words it is necessary to look at the reality of the situation, not how the agreement was labelled. The effect of the decision was to give occupant of the premises the rights of a sitting tenant.

The distinction became less crucial after 15 January 1989 when, after a quarter of a century, it again became possible for landlords to grant short residential tenancies guaranteeing their right to take back the property so long as certain documentary formalities were complied with. The Rent Act 1977 was superseded by the Housing Act 1988 for new residential tenancies granted after that date — though the rights of existing sitting tenants were preserved. The short term lettings regime was further relaxed when Housing Act 1996, provided that almost all new private lettings, at a market rent, are automatically assured-shorthold unless the landlord has previously elected to give the tenant lifetime security.

However this book is not primarily concerned with short term residential tenancies at a market rent, but with long term residential ground leases granted at a full market premium (or price) and reserving a small annual ground rent.

At the time of writing, almost all flats and maisonettes in owner/ occupation are held on ground leases. It is a workable arrangement which should ensure that the building is maintained and insured; that the cost of such maintenance and insurance is shared fairly between the leaseholders and that everyone abides by a mutual set of rules so that, as far as possible, each leaseholder can enjoy their residential unit without disturbance or unreasonable interference by other leaseholders. A freehold flat is currently unusual and would be unattractive to mortgage lenders.

Having said this the Commonhold and Leasehold Reform Act 2002 has now paved the way for a new type of freehold which developers can choose to use for flats and maisonettes, instead of selling them on lease. The new type of freehold will be termed a "commonhold" and will contain mutually enforceable rights and obligations corresponding to those contained in a traditional lease. As such it should satisfy the concerns of mortgage lenders. The new alternative will be voluntary and, for the foreseeable future, is likely to run alongside the continued grant of traditional residential leases.

The structure of a residential lease

If the reader was to compare a modern residential lease with one signed 150 years ago, they would appear to have little visibly in common.

The earlier lease would comprise a mass of shakespearean language copperplated across parchment. Its modern counterpart would be substantially longer, printed in modern legalease and on A4 paper.

But while the appearance and terminology of the later document may bear no resemblance to the earlier, the actual legal structure has scarcely changed. Both will have followed a tried and tested formula which has evolved over the centuries. It is the format, which all property professionals have grown used to. It ensures that important contractual provisions relating to responsibility to repairs, insurance, alterations to the property, permitted use and relevant financial obligations can be found quickly and without having to study the entire document. The basic structure and order of a conventional lease is as follows: particulars and definitions; the demise; tenants covenants; landlords covenants; matters agreed and declared; schedules and ending up with the attestation (or signature) clauses. Each section of a conventional lease is explained later in this chapter.

It is also convention that each lease is not signed in duplicate by all parties but as an "original" and "counterpart". Although the main body of each document is identical, the "original" will be executed only by the landlord — and the 'counterpart' will be executed only by the tenant. On completion of the lease, each party will exchange their signed duplicate of the lease, so that the lessee ends up with the landlord's signed original — and the landlord ends up holding the tenant's signed counterpart.However, the fact that lawyers have prepared leases in this way for centuries does not mean that the process will not undergo radical change in the foreseeable future.

Moves towards electronic conveyancing mean that future leases may exist only in cyberspace. Paper deeds and documents will be replaced by their "virtual" equivalents. Legislation enabling this to happen is already contained in part 7 of the Land Registration Act 2002. What is currently missing is technology sufficiently developed to provide a system of electronic conveyancing, which is secure against fraud or computer malfunction. Another likely reform, which could take place at any time, is the standardisation of all new leases within a prescribed Land Registry format, dictated by regulations. But however "virtual" or "standardised", residential leases become, the fundamental provisions must remain the same.

Stamp duty and stamp duty land tax

Before 1 December 2003 any residential lease granted for a premium above the lowest stamp duty threshhold (currently £120,000) would be expected to bear a series of red stamps showing that the required Stamp Duty had been paid. Even for leases granted at less than this minimum threshold there would have been some duty payable on the ground rents, no matter how small. Each counterpart lease was also stamped with a nominal £5 duty.

With the replacement of Stamp Duty by the Stamp Duty Land Tax (SDLT), those red stamps have gone (as has duty on counterparts). The Inland Revenue will instead issue a certificate confirming that the relevant tax has been paid. SDLT is also paid whenever a freehold or lease is sold and duty will be calculated on a rising scale according to the price paid for the new lease or transfer. From £120,001 to £250,000, SDLT is payable at 1%. From £250,001 to £500,000, SDLT is payable at 3%. Above £500,000 it is payable at 4%. SDLT is also payable on any ground rent or market rent with a capitalised value of more than £120,000 and there is a complex statutory formula for calculating this.

Without evidence that the required SDLT has been paid, a property transaction cannot be registered at HM Land Registry. Without such registration, legal title will not transfer to the prospective lessee or purchaser. SDLT must be paid within 30 days from the "effective date" of a transaction. This time-limit is backed up with a rising scale of penalties for late payment.

The rules relating to SDLT are complex are continually being updated as the soon system settles in. Full information including a "lease duty calculator" can be found on the Inland Revenue website. Some areas within the UK are designated by postcode as "disadvantaged". In those areas SDLT does not begin to be payable on a residential property unless the premium or purchase price is at least £150,000.

Leases and title registration

An owner's title to a leasehold property comprises two components:

1. the lease itself — which sets out the names of the original parties to the lease, the extent of the property leased and the terms and conditions on which it is leased and
2. the information about current ownership of the lease, which is held centrally by HM Land Registry.

Until 12 October 2003 the Land Registry issued certificates of title to property owners and mortgagees confirming their proprietary interest in the property. If the property were not in mortgage a land certificate would be issued. If it were in mortgage a charge certificate would be issued. As part of its move towards computerised conveyancing, land and charge certificates were abolished on 13 October 2003 although many are still in existence. In its place the Land Registry now issues title information documents. These are not deeds in themselves and they have no intrinsic value. They provide up to date information about current ownership and any financial charges.

Each registered title is allocated with its own title number. That number will remain the same, no matter how many times the property is transferred as a single unit. The register itself is divided into the three parts.

- *The property register* — which provides a brief description of the leased premises cross referenced to a title plan showing the location of the leased-premises within the building within which it is situated and in relation to adjoining streets. The property register will also state the period of the lease and the names of the original lessee and lesser.
- *The proprietorship register* which states first the class (or quality) of the leasehold title. The best title is "absolute leasehold", which signifies that the Land Registry has checked and guarantees the landlord's freehold title and any superior leasehold titles as well as the leasehold title created out of it. "Good leasehold", means that the freehold title has not been warranted by the Land Registry. However, in some cases a good leasehold title will be acceptable to mortgage lenders. The proprietorship register will also state the full name and address of the current owner of the lease, as at the date title was last transferred. Finally, the proprietorship register will state any restrictions on the right of the current registered proprietor to sell or mortgage their interests.
- *The charges register* — will disclose any freehold title restrictions or other matters affecting the freehold which also are significant to the leasehold interests. It will also disclose third party interests affecting only the leasehold title (such as a registerable sublease). It will also list any mortgages affecting the leasehold title, giving the date of the mortgage and the name of the lender.

The Land Registration Act 2002 now requires any lease of more than seven years to be registered and given its own title number. Compulsory registration requires the application for registration to be made within two months from completion of the lease. In fact, registration must be made twice: once to create the new leasehold title — and second as a notice against the landlord's title (if registered), to alert anyone buying or lending against the landlord's interests to the fact that there is a lease (or leases) in existence. Compulsory registration also applies if an existing unregistered lease with more than seven years outstanding is transferred to someone else or mortgaged.

Before 13 October 2003 the threshold for compulsory first registration was 21 years. In practice most residential ground leases, being for terms of 99 years or more, will have been long registered. Though undoubtedly there are some long residential leases which were signed before compulsory registration took effect in that locality, and which has not been transferred or mortgaged since.

Compulsory first registration spread piecemeal across England and Wales over the course of a century and only became complete as recently as 1990. But if there are few unregistered leaseholds, there will be more unregistered landlords' titles as the grant of a lease does not trigger a requirement for registration of the freehold title out of which it was created. A landlord's title, which is registered, will be cross-referred to in the Property Register of the relevant leasehold title.

HM Land Registry operates out of a series of district offices, with specified areas of England and Wales allocated to each.

The Land Registry is also "open", meaning that anyone can obtain a copy of any registered title and any title-plan, on completing the correct form and sending it to the appropriate district office with a small fee (currently £4 for a copy of the register and £4 for a copy of the title plan).

These title registers provide an essential source of information for any leaseholder or (group of leaseholders) wishing to exercise any of the rights set out in later chapters. It enables them, at small expense, to obtain full information about the identity of the ground landlord and other leaseholders, including relevant information about their titles, including details of mortgage lenders.

Mortgageability of leases

The main mortgage lenders jointly publish the *CML Lenders Handbook for England and Wales*, which is available only on-line and periodically updated. It is written for property lawyers and set out lenders' expectations as regards the title checks and associated matters which lawyers acting on their behalf must make to ensure that title to the property is "good and marketable" and that the mortgage security is sufficiently protected. A lawyer who fails to comply with any CML requirement will be exposed to a damage claim, if the mortgage lender loses money as a result. It follows that a residential lease which does not comply with the terms *CML Lenders Handbook* will be neither mortgageable or saleable — at least not without expensive title-indemnity insurance or other remedial action.

A copy of the Handbook can be downloaded from the Council of Mortgage Lenders website *www.cml.org.uk*. Part I of the *Handbook* contains general guidance applying to all mortgage lenders — while Part II deals with the specific additional requirements of particular mortgage lenders. That part of the *Handbook* relating specifically to leases is set out in section 5.2 of the 2003 edition and its current requirements are explained later in this chapter.

The remainder of this chapter looks at the structure of a typical residential ground lease. As each lease is different it would be inappropriate to follow a typical lease clause by clause. Instead the author has grouped the content of such leases under subject headings — starting with the most important. Readers may find it helpful to have an actual copy lease before them with which to cross refer. Restrictions on alterations, use, under letting, and transfer are dealt with separately in Chapter 4.

The demise

This is the most important clause in any lease. It is the clause on which everything else hangs. It is the clause which transfers legal possession of the residential unit to the lessee (and their successors) for the period of the lease in return for the capital premium, ground-rent and compliance with all other leasehold obligations. It is found towards the beginning of the lease after essential words and phrases have been defined. A typical "demise" might reads as follows:

> In consideration of the Premium now paid by the Lessee to the Lessor (the receipt whereof the Lessor hereby acknowledges) and of the rents and covenants hereinafter reserved and contained and on the part of the Lessee to be performed and observed the Lessor with full title guarantee HEREBY DEMISES unto the Lessee ALL THAT the Demised Premises together with the Rights but subject to the Reservations TO HOLD the same unto the Lessee for the Term YIELDING AND PAYING the Rent by equal half-yearly payments on the Payment Dates free of all deductions the first payment being an apportioned part of the said rent calculated from the date hereof to the next Payment Date to be made on execution hereof AND BY WAY OF FURTHER RENT the Service Charge to be paid in accordance with the provisions of the [third] schedule hereto.

Terms such as the premium, rents, demised premises, term, service charge will be defined elsewhere in the lease, if not within the clause itself. The term "with full title guarantee" refers to the "Implied Covenants of Title" given by a vendor or lessor to a purchaser or lessee under Part I of the Law of Property (Miscellaneous Provisions) Act 1994.

"Full guarantee" involves warranties by the lessor to the lessee that the lessor legally has the right to grant the lease and will do all that he can to confer legal title to the lessee; that there are no undisclosed third party interests affecting the property; and that if the landlord is himself a lessee, and not a freeholder, the landlord's lease is still existing and that nothing has happened which may render it liable to termination. Where property is transferred or leased with limited title guarantee, the lessor or transferor does not warrant the absence of undisclosed third party interests, save such third-party interests as he has created himself or of which he is aware since the property was last sold. Sales with limited title guarantee are usually made by vendors or lessors who have not personally occupied or managed the property which they are selling or leasing. Examples might include mortgage repossessions or sale by bankruptcy trustees or executors of a deceased person's estate.

The premises or demised premises

The second most crucial clause is the extent of the premises to which the lease relates.

Not everything within the flat or maisonette would belong to the lessee. External walls as well as internal load bearing walls (save for the internal plaster surfaces of such walls) and floor joists or slabs will usually be excluded. Such items form part of the main structure of the building and any interference with them could affect other leaseholders or the general stability of the building within which the flats or maisonettes are situated. As such, ownership and maintenance of such items must remain with the ground landlord, with the cost of such maintenance being shared between the relevant leaseholders. Thus, leaseholders who affix satellite dishes on the outside walls of their units may technically be trespassing on a part of the building which does not belong to them.

In general a brief description of the premises (or demised premises) will appear among the particulars at the beginning of the lease. This will cross refer to a schedule describing in the minutest detail which parts of the unit are "included" within the lease and which parts are "excluded". That schedule will in turn cross-refer to a floor plan of a leased property, which is also bound into the lease. As well as a detailed description of the premises themselves, there will also be schedules of "rights" and "reservations".

A flat or maisonette does not exist in isolation. Lessees and their visitors will need to enter other parts of the building to get to and from the residential unit. Water, gas, electricity and other utility services serving the unit will also travel through pipes and wires situated in other parts of the building — as will sewers and drains. There may also be gardens and other communal areas for everyone to enjoy.

Therefore in addition to the unit itself, the lease will also need to grant the "rights" to walk along the corridors and stairways of the building, use lifts, obtain utility services and expel waste as well as use communal facilities, such as gardens. Those "rights" will either be set out as separate schedule or in a separate part of the same schedule as the premises-description. Then there are the "reservations".

These are the rights which the landlord and other leaseholders retain in relation to the leased premises. For instance within the premises there may be pipes, wires, cables, drains and sewers within the flat or maisonette which serve or are shared with other flats or maisonettes — and which will need to be the subject of reservations. Again, the reservations will be set out in a separate schedule or in a different part of the same schedule. To a large extent the rights and corresponding reservations will mirror each other.

Less obvious but equally important are the rights of "support and shelter" which exist both as a right and reservation and are essential for the stability of the building and each unit within it. The covenants are there because each flat (above ground level) will be physically "supported" by those flats and other parts of the building situated below it. Equally all the flats will be dependent on the existence of the flats and roof above it for "shelter". This mutual right and reservation reflects the fact that each unit is dependent, for its structural stability, on the continued existence of every other part of the building.

The foundations and roof of the building (though not necessarily the loft-space), as well as the communal areas, will remain within the ownership and responsibility of the ground landlord, subject to the rights of support and shelter as are referred to above.

Financial provisions

A conventional residential ground lease will contain the following financial provisions.

- The *premium* (or purchase price) paid by the original lessee to the ground landlord when the lease was granted. This simply records the amount originally paid and has no other significance.
- The *ground rent*. This may be nominal (eg a peppercorn or say £5 per year) or it may be substantial. It is paid by the current lessee to the current lessor throughout the term of the lease. Many leases provide for fixed increases in the ground-rent to compensate for inflation, after a certain number of years have elapsed. In some cases those staged increases may be linked to the current value of the premises if sold on the open market (eg 0.5%). One of the benefits of a ground rent is that it keeps ground landlords and their lessees in touch with each other. It is when the effects of inflation erode the value of a fixed ground-rent to an insignificant amount that many landlords do not bother to collect it, that contact between leaseholders and their landlord becomes lost.
- *Service charges* — which represent each leaseholder's contribution toward costs incurred (or to be incurred) by the ground landlord in insuring, maintaining, repairing and managing the building. This is usually fixed, for each leaseholder, as a percentage of the total estimated cost to the landlord of managing the building — or loosely described in some leases as a "fair proportion". The list of landlord-services for which lessees can be required to contribute will conventionally be set out in a separate schedule to the lease and may include a "sweeper" entitling the landlord to add further items to the list as circumstances require. Service charges will commonly require payment on account towards estimated service costs — with provision for adjustment (or carrying forward) once the actual costs are known.
- *Tenant's outgoings* — being council tax, utility charges and related expenditure affecting only the particular unit for which individual leaseholders will take responsibility directly.

The right of individual leaseholders to be consulted on major items of service charge expenditure and to challenge unfair service charges is explained in Chapter 5.

Insurance

The cost of annual buildings insurance is dealt with through the service charge provisions of the lease. The actual placing of the insurance contract is a landlord's responsibility and there should be an express obligation to that effect within the landlord's "covenants". A single insurance contract should be taken out in respect of the entire building and be sufficient to cover the costs of full reinstatement.

In relation to such insurance, the landlord is under two additional contractual obligations:

1. in the event of damage or destruction by an insured risk, to lay out all resulting insurance proceeds in reinstating the building and
2. to produce evidence of such insurance on request to each individual lessee.

A landlord's insurance obligation may or may not list out the risks which must covered in any policy of insurance. The *CML Lenders Handbook* requires such insurance to cover: fire, lightning, aircraft, explosions, earthquake, storm, flood, escape of water or oil, riots, malicious damage, theft or attempted theft, falling trees and branches and aerials, subsidence, heave, landslips, collisions, accidental damage to underground services, professional fees, demolition and site clearance costs, and public liability to anyone else.

Where a separate management company is made party to the lease, that company may be responsible for the building's insurance. Where a house is leased in its entirety it is more likely that the leaseholder is made personally responsible for their own building's insurance in respect of their property, with similar provisions as to laying out of insurance proceeds in the event of damage or destruction. Such clauses would also require the tenant's insurer to be nominated or approved by the lessor, with an additional requirement that such insurance be in the joint names of the lessor and the lessee. Note also section 164 of the Commonhold and Leasehold Reform Act 2002 (which only applies to houses) and which states that a leaseholder is no longer required to insure with a company nominated or approved by the landlord so long as the leaseholder has made alternative arrangements. See also the Leasehold Houses (Notice of Insurance Cover) (England) Regulations 2004.

Maintenance and redecoration

In general, each individual lessee will be directly responsible for the internal maintenance and redecoration of their own unit. This includes responsibility for maintenance of any drains, pipes, wires or other apparatus, which serve the unit *exclusively*. As well as the general tenants "repair" clause, there will be a separate tenant's obligation requiring repair or decoration at specified intervals.

Responsibility for the exterior, main structure, foundations, roof and communal areas will be on the landlord, on the basis that such costs will be recharged to individual leaseholders through the annual service charge.

Provisions governing behaviour

All residential leases will contain a list of provisions designed to prevent anti-social behaviour by individual lessees. Examples are provisions to prevent: illegal or immoral activities, noise nuisance, the storage of dangerous substances, keeping of animals, hanging washing, inconsiderate parking, sales or auctions; and to prevent other types of nuisance or annoyance.

As each lease represents a contract only between each individual lessee and the ground landlord, lessees cannot enforce these provisions against each other directly. Only the landlord can enforce them.

The *CML Lenders Handbook* requires every lease to contain adequate provisions for the enforcement of such obligations by the landlord or management company at the request of the tenant. Most leases will therefore require the landlord to enforce such restrictions at the request and expense of the leaseholder requesting such enforcement.

The forfeiture clause

The forfeiture clause provides the "teeth" of any lease. It is the ultimate sanction against non-compliance. It means that any leaseholder who is in persistent and serious non-compliance of the lease can have that lease taken away by the ground landlord without compensation. Without a forfeiture clause no leaseholder would have any legal incentive to comply with any restriction in the lease — possibly creating a situation which is not only intolerable for the ground landlord but more likely intolerable for other leaseholders who are forced to live with anti-social behaviour.

The forfeiture clause is conventionally one of the last clauses in the lease (before the schedules). It is conventionally written in harsh terms. For example:

> If the rents reserved or any part thereof shall be unpaid for twenty days after becoming payable or if the Lessee shall not observe and perform all and every other covenant, condition, restriction, regulation, obligation and agreement on the part of the Lessee herein contained then and in every such case it shall be lawful for the Lessor or anyone authorised by the Lessor to re-enter the Demised Premises or any part thereof in the name of whole and to repossess the same and thereupon the term hereby created shall cease and determine but without prejudice to any right of action or remedy of the Lessor in respect of any antecedent breach of any of the covenants on the part of the Lessee herein contained.

Fortunately the reality is not so severe. It would be unfair on a leaseholder (as well as on any mortgagee) to allow a ground landlord to snatch away a valuable capital asset on pretext of some isolated and trivial breach. Although this is not to say that a residential lease can never be forfeited where there is an obvious and continuing breach. In this respect forfeiture has traditionally been more effective as a threat than as a reality. There are also the following legal hurdles which any ground-landlord wishing to forfeit must first overcome.

The need for a court order

In most cases where a property is occupied residentially, the Protection from Eviction Act 1977 will require any proposed eviction to be sanctioned by the court before it can take effect. While there are exceptions for accommodation shared with the landlord and for holiday or entirely gratuitous lettings, any occupied residential flat held under a long lease will qualify for the full protection of the law under section 2 of the Act. It follows that any forced entry by a landlord without a court order will count as an unlawful eviction under section 1 of the Act, which is a criminal offence punishable by imprisonment. Unlawful eviction can also form the basis of a civil claim under section 27 of the Housing Act 1988.

Where unlawful eviction is proved, section 28 of the 1988 Act empowers a court to grant punitive damages to the evicted tenant sufficient to cancel out any profit, which the landlord might have hoped to make from his unlawful actions. Use of actual (or threatened) violence by a landlord to repossess residential premises from a leaseholder (or former leaseholder) physically within the premises is also made a criminal offence by section 6 of the Criminal Law Act 1977.

But it should not be taken for granted that a court order will always be required before residential premises can be repossessed. It is still conceivable that a leaseholder who goes abroad for an extended stay without making adequate arrangements for payment of rent and service charges whilst he is away, could return to discover that his lease has been forfeited and that his flat is now occupied by strangers.

Relief from forfeiture

The courts have always been willing to grant a tenant relief where a lease has been forfeited on the grounds of rent arrears and the tenant subsequently pays off those arrears within a reasonable time and reimburses any associated out of pocket expenses incurred by the landlord as a result of that default. The granting of such relief means that both the lease and the tenant are reinstated. In circumstances where a lease has been forfeited without a court order, the tenant seeking such relief must apply to the court as soon as possible. But although the law's willingness to grant relief against forfeiture has always existed in relation to rent arrears, it was not until the Conveyancing and Law of Property Act 1881 that relief was extended to other forms of breach (see below). When a landlord applies to a county court for forfeiture of (any) lease on grounds of rent arrears, section 138 of the County Courts Act 1984 requires the court to grant relief to the tenant if those arrears (and the landlord's costs in bringing the proceedings) are paid *into court* at least five clear days before the scheduled hearing date.

Even if such payment has not been made before the hearing, the court must still give the leaseholder at least another four weeks to make a payment into court bringing the rent up date (and discharging any associated costs liability) before forfeiture becomes final.

The need to serve notice under section 146 Law of Property Act 1925

Subject to compliance with the other various statutory requirements relating to rent demands, a ground landlord can apply immediately to the court for forfeiture if ground rent is significantly in arrears. The issue of legal process will itself represent notional "forfeiture", subject to the leaseholder's right to apply for relief (see above). Where forfeiture is to be sought on any other grounds (such as anti-social behaviour or unauthorised alterations) the landlord must first serve notice on the tenant under section 146(1) of the Law of Property Act 1925:

(a) specifying the particular breach complained of and
(b) if the breach is capable of remedy, requiring the lessee to remedy it and
(c) in any case, requiring the lessee to make financial compensation for the breach.

The effect of the landlord's notice is to give the leaseholder one final opportunity to rectify the breach and compensate the landlord before action is taken to forfeit the lease.

If the leaseholder complies with the notice, that is the end of the matter. If the leaseholder does not comply with the notice, within a reasonable time, action may be taken to forfeit the lease unless the court can be persuaded to grant relief against forfeiture — "having regard to proceedings and conduct of the parties and to all other circumstances". In the landmark case of *Billson* v *Residential Apartments Ltd* [1992] 1 EGLR 43, the House of Lords confirmed the general rule that a tenant's right to apply for relief remains preserved, even if re-entry has been lawfully effected without a court order. The decision of Deputy Judge Jarvis QC in *Mount Cook Land Ltd* v *Hartley* [2000] EGCS 26, illustrates how far the courts will go to prevent confiscation of a valuable leasehold asset.

Lessees had wrongfully sublet premises without the landlords consent as required by the terms of their lease. One of the leases had 34 years left to run at an annual ground rent of £300. The same premises had a capital value to the leaseholder of £300,000 and an open market rental value of £42,250 per year.

The landlord served notices under section 146 threatening forfeiture for breach of covenant. The leaseholders applied to their landlord for retrospective consent to the underlettings. However, the landlord declined to consider the requests for consent. Instead their solicitors wrote to the leaseholders saying that the contravention of the lease-terms was incapable of remedy and that the landlord was consequently regarding the leases at an end. Possession proceedings followed and the leaseholders sought relief from forfeiture. Such relief was granted, the judge saying:

> This is a question of proportionality, weighing up the advantage to the landlord of recovering possession as against the loss to the tenant. If I were to refuse the relief that would be a windfall that the landlord would receive and I have to weigh up whether that kind of loss is proportionate to the breach that has occurred, and whether it would represent a fair result. In my judgement, looking at a question of proportionality, therefore, the amount of damage sustained by the landlord compared to the damage to the tenant, if I did not grant relief as against the benefit to the landlord, is, it seems to me, overwhelmingly against the landlord. The landlord's gain would be totally out of proportion to what has in fact occurred.

He added:

> It does seem to me that the landlord saw an opportunity to gain possession of a valuable property and was set on that course.

He also agreed with the tenant's lawyer that as the unauthorised sub-lettings were at a full market rent it was difficult to see upon what basis the landlord could have refused to consent.

In circumstances where notice has been served under section146(1) threatening forfeiture, section 146(2) gives lessees a statutory right to apply to the court for relief against forfeiture — which the court may either grant or refuse, having regard to the nature of the proceedings, the conduct of the parties and to all other circumstances. Such relief may be granted upon such terms as to costs, expenses, damages, penalty, the granting of an injunction, or otherwise, as the court deems appropriate. Note also section 147 of the 1925 Act, which allows a lessee to apply to the court for relief when notice has been served under section 146 relating purely to the failure to comply with an obligation relating to internal decorative repairs.

If the court is satisfied that the section 146 notice is unreasonable, it may wholly or partly relieve the lessee from liability for such internal decorative repairs. However, even in relation to internal decorative repairs, lessees cannot be relieved from any obligation in any of the following circumstances:

- where there has been an express agreement on the part of the lessee to put the property into a decorative state of repair, and that agreement has never been performed

- to work which is necessary or proper for putting or keeping the property in a sanitary condition or for the maintenance or preservation of the structure
- so far as it relates to any statutory liability to keep the property reasonably fit for human habitation or
- to any agreement to yield up the property in a specified state of repair at the end of the lease term.

To prevent landlords issuing section 146 notices inappropriately to bully leaseholders in accepting unreasonable demands, section 168 of the Commonhold and Leasehold Reform Act 2002 now prohibits a landlord from serving such notices on long residential leaseholders unless such breach has either been admitted by the tenant or proved by the landlord in proceedings before any court, leasehold valuation tribunal or arbitrator (see Chapter 14).

The Leasehold Property (Repairs) Act 1938

The Leasehold Property (Repairs) Act 1938 states that where a section 146 notice is served in relation to a tenant's alleged failure to keep the property in repair during the currency of the lease, the lessee may within 28 days serve a counternotice claiming the benefit this Act. Where the counternotice has been served, the lessor can take no action to forfeit the lease without prior leave from the court. Such leave cannot then be given unless the lessor proves any one of the following:

- that the breach must be immediately remedied to prevent a reduction in the value of the lessors' interest or that the value has already been substantially reduced by the breach
- that the breach must be immediately remedied to comply with any relevant legislative requirements
- in a case where the lessor is not personally in occupation, that the immediate remedying of the breach is required in the interests of the occupant of the premises
- that the breach can be immediately remedied at a relatively small expense compared with the much greater expense that might be occasioned if the work is postponed
- that there are special circumstances making it just and fair that leave be given.

Waiver of breach

A landlord cannot take steps to forfeit a lease for an alleged breach of a lessee's non-monetary obligation and at the same time continue to demand or accept future rent. The landlord must decide one way or the other. Demanding or accepting future rent in such circumstances may amount in law to a "waiver" (or acceptance) by the landlord of the lessee's wrongful conduct and so neutralise the contravention as a ground for forfeiture. A waiver can only apply to a past "one-off" breach, such as unauthorised alterations or an unlawful assignment or subletting. It cannot apply to rent or service charge arrears. It will only apply when it can be proved that the landlord was aware of the breach before demanding or accepting rent.

Often the first warning a leaseholder gets of a pending forfeiture is when rental payments are refused or rent-cheques returned by the landlord. The Court of Appeal decision in *Chrisdell Ltd* v *Johnson* [1987] 2 EGLR 123 illustrates how the courts approach waiver.

A Rent Act protected residential lease for three years contained a clause prohibiting assignment. Notwithstanding the prohibition, the named tenant Johnson, went away leaving Mrs Tickner in occupation. His written explanation to the landlord was that Mrs Tickner was a housekeeper whom he had asked to look after the premises while he was away. This was in spite of the fact that Mrs Tickner had paid him a capital sum of £3,000 plus an additional sum of £35 per month. Nevertheless the landlords accepted his explanation and continued to demand and receive rent under the lease.

Shortly afterwards, the landlord sold its interest to Chrisdell Ltd, who became Johnson's new landlord. Chrisdell issued proceedings for possession based on Johnson's wrongful assignment, subletting or other parting with possession of the premises to Mrs Tickner. Mrs Tickner responded by claiming that Chrisdell was bound by the previous landlords "waiver" of the breach. Had Mrs Tickner succeeded in her argument, she could have remained in the premises for life as a Rent Act protected tenant. The court disagreed.

The former landlord had not "waived" the contravention because they had relied on the previous explanation put forward by Johnson and his solicitors. It followed that the new landlord, Chrisdell, was able to rely on the breach as providing grounds for possession of the maisonette.

Sections 47 and 48 Landlord and Tenant Act 1987

Section 47 of the 1987 Act requires any monetary demand given by a landlord in writing to a tenant in premises to contain the name and address of the landlord and (if that address is not in England and Wales) an address in England and Wales at which notices (including proceedings) may be served on the landlord by the tenant. In the absence of such information, any part of the financial demand which comprises a service charge (or an administration charge) is not regarded as being due until such time as the information is given to the tenant by notice from the landlord.

Section 48 imposes a general requirement on landlords to provide tenants with an address in England and Wales at which notices (including proceedings) may be served on the landlord by the tenant. Again failure to provide such written information means that any rent, service charges or administration charges are not regarded as being due until that information is provided. The effect of sections 47 and 48 is to provide tenants with a technical defence against either forfeiture or any other proceedings for recovery until the statutory requirements are complied with. But once such information has been provided, the landlord will be entitled to back payment for all sums previously owing. The initial information relating to the landlord's name and address may be contained within the original lease or tenancy agreement itself. New notification will have to be promptly served on each leaseholder if the landlord's interest is transferred to a third party.

These notification requirements are now extended by section 166 of the 2002 Act which states that a leaseholder is not required to make a payment of rent under a lease unless the landlord has first given the tenant notice relating to the payment. This notice must specify the amount of the payment and the date by which it must be paid. The leaseholder must be given not less than 30 days and not more than 60 days after issue of the notice to pay the ground rent. The form of the notice must conform to the Landlord and Tenant (Notice of Rent) (England) Regulations 2004.

Forfeiture for financial arrears

Section 167 of the Commonhold and Leasehold Reform Act 2002 prevents a ground landlord exercising forfeiture for financial arrears unless the unpaid amount either exceeds the prescribed sum or has remained outstanding longer than the prescribed period. Under the

Rights of Re-entry and Forfeiture (Prescribed Sum and Period) (England) Regulations 2004, the prescribed sum is set at £350 and the prescribed period is set at three years. Note also that landlords cannot increase small arrears above the statutory threshold by adding a "default charge".

Forfeiture and disputed service charges

Since 1996 landlords have been unable to commence forfeiture of a residential lease for unpaid service charges unless the amount of such default has either been admitted by the leaseholder or proved by the landlord in proceedings before a court, a leasehold valuation tribunal or in arbitration proceedings (where allowed). The current law is contained in section 81 of the Housing Act 1996 (as amended by section 170 of the 2002 Act) which prevents a landlord from exercising forfeiture for alleged service or administration charge arrears unless either that liability has been previously established before a court, leasehold valuation tribunal (LVT) or arbitration proceedings, or that the tenant has previously admitted his liability in this respect. There is a further postponement of 14 days of the landlord's right to institute forfeiture after proceedings are finally concluded.

When forfeiture is inevitable

Forfeiture is of course only one option for landlords faced with leaseholder default. Other options, for financial arrears, include the institution of a simple money claim through the civil courts or insolvency proceedings. In some cases, landlords faced with small arrears may write directly to the lessee's mortgagee. In such circumstances, lenders may settle the arrears themselves, to protect their security, before adding it to the mortgage debt. For anti-social behaviour or other threatened physical breaches, injunctions may provide a remedy.

While recent statutory reforms prevent the forfeiture being used or threatened inappropriately, it is a mistake to think that landlords can no longer use it to deal with serious or persistent lease contraventions. The situation when forfeiture is most likely to arise in practice is when a leaseholder is no longer able to meet legitimate service charge obligations. Arrears will then spiral and, if no other solution is found, the leaseholder could be faced with the loss of his lease.

Once a ground landlord has cleared all statutory hurdles, a judge can only prevent forfeiture from taking place if the leaseholder can demonstrate a reasonable prospect of being able to clear the arrears (and associated costs) within a reasonable time. If the leaseholder has insufficient income, the stark reality is that he may be unable to avoid forfeiture.

In such circumstances it is better that the leaseholder seeks a solution with the ground landlord than waits for the inevitable. Even if the lease-holder's overtures to the landlord are rebuffed, the fact that such overtures have been made — and are on record — may still carry weight with a county court judge. Once forfeiture takes effect, all financial equity in the property will pass from the leaseholder to the lessor.

It is even more important that leaseholders make their mortgage lenders aware of the situation — as their financial security will also be lost if the lease becomes forfeit. If this happens, the mortgage lender can still sue the former leaseholder for the now unsecured debt. In such circumstances it may be mutually better for the mortgagee to repossess the unit and exercise its own power of sale — thus enabling any rent/service charge arrears to be paid out of the sale proceeds. Any surplus (after deduction of the mortgage debt and associated costs) will then belong to the former leaseholder.

A leaseholder faced with this difficult situation should also consider a negotiated sale of the unit at the best price which can be achieved — in the expectation that he can come away with something after all arrears, mortgage-lenders and other liabilities have been repaid.

The covenant for quiet enjoyment

The tenants paying the rent and observing the terms and conditions on their part contained in this lease shall be entitled to quiet enjoyment of the premises during the term hereby granted free from interference by the landlord or anyone claiming through or under it.

This provision, or words to like effect is included in every lease whether short or long, residential or commercial. If the clause is not set out expressly, it will be implied by law, and has done so for centuries.

It is traditionally known as the "covenant for quiet enjoyment". It means that, having granted the lease, the landlord will not thereafter do anything, or permit anything to be done which will undermine the quality of the tenant's occupation of the premises.

It's meaning is entirely technical. It does not guarantee "peace and quiet". Neither does it impose any warranty on the landlord as regards the state and condition of the premises. The effect of the covenant is explained in two cases which, by coincidence, both involve the London Borough of Southwark.

In *Southwark London Borough Council* v *Mills* [1999] 3 EGLR 35 the House of Lords was required to define the meaning of this covenant.

Mrs Tracey Tanner was a local authority "secure" tenant living in a block of flats on Herne Hill. Although her neighbours were not unreasonably noisy, the flats had no sound insulation, and she could hear not only neighbours televisions and their babies crying but their coming and going, their cooking and cleaning, their quarrels and their love making. That lack of privacy caused her hypertension and distress.

In accordance with the terms of her tenancy agreement, she commenced arbitration proceedings against Southwark Council. The arbitration tribunal ordered the council to install soundproofing and that decision was upheld by the High Court. Her tenancy agreement did not contain any landlord's warranty that the flat was sound insulated or otherwise fit to live in. Nor was such warranty implied by law. Mrs Tanner sought instead to invoke the covenant for quiet enjoyment contained in her tenancy agreement. The issue for the House of Lords was whether the matters the subject of her complaint fell within the remit of that covenant. The Law Lords ruled that they did not.

What Lord Hoffmann said about the covenant can be summarised as follows:

> The covenant for quiet enjoyment is a covenant that the tenant's lawful possession of the land will not be substantially interfered with by the acts of the lessor or those claiming under him. Two points about the covenant should be noticed. Firstly there must be a substantial interference with the tenant's possession. This means his ability to use it in the ordinary lawful way. The covenant cannot be elevated into a warranty that the premises are fit for some special purpose ... On the other hand, it is a question of fact and degree whether the tenant's ordinary use of the premises has been substantially interfered with.
>
> However I do not see why, in principle, regular excessive noise cannot constitute a substantial interference with the ordinary enjoyment of the premises. The fact that the tenants complained of noise is not therefore a reason why their actions should fail.
>
> There is however another feature of the covenant, which presents tenants with a much greater difficulty. It is prospective in its nature. It is a covenant

that the tenant's lawful possession will not [in the future] be interfered with by the landlord or anyone claiming under him. The covenant does not apply to things done before the grant of the tenancy, even though they may have continuing consequences for the tenant.

In the grant of a tenancy it is fundamental to the common understanding of the parties, objectively determined, that the landlord gives no implied warranty as to the condition or fitness of the premises. [so lessee beware]. It would be entirely inconsistent with this common understanding if the covenant for quiet enjoyment created liability for disturbance or inconvenience or any other damage attributable to the condition of the premises.

The Law Lord's judgment in *Southwark* v *Mills* provides a textbook summary of quiet enjoyment, as it also reviews earlier judicial decisions on the issues. But it was not the end of the matter as the issue came again to the Court of Appeal less than three years later in *Southwark London Borough Council* v *Long* [2002] 3 EGLR 37.

Ms Long was a secure tenant living at 5 Townsend House, Bermondsey. Her flat was next to a small utility room housing a large paladin bin into which ran a rubbish chute. Upstairs tenants could either put rubbish in the chute by opening a hopper door on one of the upper landings or put rubbish in the bin directly. The chute was inadequate to take the amount of rubbish disposed of and other tenants frequently left their rubbish behind it when the bin was full. As a result the doors to the bin cabin were not kept closed. She also complained of considerable noise from the rubbish chute, as tenants would bang the doors on the hoppers to force their rubbish down the chute. She also complained of smells and maggot infestations.

She sued the council on several grounds, one of which was that the council was in breach of its covenant for quiet enjoyment. She succeeded in this argument before Judge Goldstein, at Central London Civil Trial Centre, who interpreted the *Mills* decision to mean that in appropriate circumstances a substantial interference with the enjoyment of premises could amount to a breach of the covenant for quiet enjoyment — it being a question of fact and degree. He awarded damages of £13,500. However the Court of Appeal ruled that Judge Goldstein was wrong in the way he interpreted the covenant for quiet enjoyment.

Arden's LJ's ruling on the point can be summarised as follows:

The design defects in the refuse collection facilities at Townsend House do not entail any breach of the covenant for Quiet Enjoyment. Nor could there be any warranty as to the adequacy of those facilities implicit in that covenant. The trial-judge made no factual-finding to the effect that the noise caused by the use of the hoppers was not contemplated in 1983 [when the tenancy began] and accordingly the noise created by tenants' use of the hoppers could not constitute a breach of the covenant. The Council did not itself leave rubbish obstructing the tenants doorway and there is no evidence that the tenant made specific complaint of this to the Council.

Although Ms Long "lost" on the covenant for quiet enjoyment, the damages award was allowed to stand as she had proved that the council was in breach of other contractual obligations relating to rubbish disposal.

The Case of the Missing Landlord

Case study
Chadwell Heath flats

Ground rent had not been collected for many years. None of the leaseholders knew who the ground landlord was or how he could be contacted. There was some old correspondence from a firm of managing agents acting for a previous landlord, but that firm was no longer in business. Each of the nine leaseholders held the residue of a 999-year lease at a fixed annual ground rent of £15.75.

In the meantime the leaseholders had collectively made their own arrangements for buildings insurance and contributed towards essential repairs, which they carried out directly. They also paid for a cleaner and someone to cut the grass. The arrangement worked until they were faced with a series of expensive repair items — some of which were urgent and others which were non-urgent but which had to be dealt with at sometime. Individual leaseholders also had different expectations as to the quality and cost of the work to be carried out and whether other improvements should be made to the building (such as a secure door entry system).

Without unanimity on these issues, the informal arrangements could no longer continue and something more formal had to be put in place. The leaseholders were also faced by a trespass on their shared car park by the drivers of motor vehicles using nearby commercial premises.

The leaseholders decided that they wanted to collectively buy out their freehold using powers contained in the Leasehold Reform Housing and Urban Development Act 1993. It was agreed that they would set up their own management company for this purpose and that decisions would be made on the basis of a majority. But to exercise their "right to enfranchise" they first had to find the existing ground landlord.

Enquiries of the Land Registry revealed that the existing ground landlord was registered as a Mr Hassan Leslie of Gants Hill. Letters addressed to him went unanswered and a recorded delivery letter was returned marked "uncollected".

Two of the leaseholders then visited the landlord's address. A new family was living there. Although current occupiers of that address were unable to provide information as to the landlord's whereabouts, they believed that he had died a few years previously and that his widow had moved to the south coast.

Armed with this information the leaseholders searched the Family Records Centre for evidence of his death and next of kin. Nothing was revealed.

However the leaseholders struck lucky during a follow-up telephone call to the local Registry of Births, Marriages and Deaths. Although the Registry had no record of a Mr Hassan Leslie, a sharp-eyed official spotted that a Mr Leslie Hassan had died several years previously.

Attention then focussed on the Probate Registry for details of any grant of representation in respect of Mr Hassan's estate. Again there was nothing. No grant of representation had been taken out by anyone.

The landlord's estate was in a legal limbo. Without a grant of representation there was no one legally responsible for Mr Hassan's estate, and no one with whom the lessees could deal.

In such circumstances, the 1993 Act laid out a procedure whereby, the court could transfer ownership from an unknown landlord to leaseholders' nominee, in return a payment into court of an amount equivalent to the value of the landlord's interest. An application was made to the local county court and an order granted. The leaseholders then applied to the leasehold valuation tribunal to assess the value of the landlord's interest. This was done and the appropriate amount paid into court to cover the value of Mr Hassan's interest plus six year's back-rent. A judge then signed the document transferring the landlord's title to the management company set up by the lessees and title was then registered at HM Land Registry. The end result is that the leaseholders now have full legal responsibility for their own affairs and can make decisions on the basis of a majority vote.

Missing landlords are more common than readers might expect. A scan of published LVT decisions will show unknown landlord cases now make up a significant proportion of LVT business. And it can present a real problem to leaseholders.

Without a traceable landlord there is no one legally responsible for the maintenance, repair, insurance or management of a block of flats. Only a landlord can lawfully act to prevent trespass by third parties on the communal areas of leased premises, such as a car park. Only a landlord can enforce payment of service charges and ensure compliance with other leasehold obligations. It is therefore not surprising that mortgage lenders will require new purchasers of flats and maisonettes to take out title indemnity insurance where a landlord cannot be traced.

Fortunately the absence of a traceable landlord does not prevent leaseholders from exercising their statutory rights to collectively buy out their freehold or take a formal transfer of management (see Chapters 6 to 11). In those circumstances the court — or an LVT — can "vest" collective ownership or management in the leaseholders' nominee. However even if the identity or whereabouts of a landlord are unknown, leaseholders must still demonstrate that they have made realistic efforts to trace the landlord and that those efforts have been unsuccessful. Those efforts may in some cases require press advertisement or even the services of an enquiry agent. But the starting point must always be HM Land Registry.

Even if there has been no contact with a landlord for many years, the likelihood is that Land Registry records will state the full name of the current landlord and his address at the date he acquired the freehold. Each leaseholder's registered title should also include the number of the landlord's title (if registered). An official copy of the landlord's registered title can then be purchased from the relevant district office using Land Registry Form OC1. Those official copies will not only state the landlord's name and last recorded address — but also disclose whether the landlord's title is itself in mortgage and provide details of the lender.

Alternatively a search of the Land Registry Index Map using Land Registry Form SIM will provide a list of all known title numbers affecting the property — including those of other leaseholders. Enquirers can then select those title numbers for which they need official copies. Of course there is no guarantee that the landlord's title is registered or, if it is, that the landlord has not moved on. But assuming that leaseholders can at least establish the name of a registered landlord, what happens next depends on whether the registered landlord is an individual or a company.

If the landlord is a company registered in England or Wales a search of the Companies House website (www.companieshouse.gov.uk) will quickly establish whether that company still exists and — if so — where it can officially be contacted. The website will also record any changes of name. Names and addresses of directors can also be purchased from Companies House at modest cost. If the company has recently been — or is in the process of being — dissolved, the website will provide the name and address of liquidators, auditors or accountants who should be contacted.

Before a company is either dissolved or struck off the Companies Register, any property assets is the name should either have been transferred out of the company or otherwise accounted for. If this is not appropriately dealt with, any residual asset belonging to the company at the time of dissolution will pass to the Crown as unclaimed property — known officially as *bona vacantia*. If it is suspected that the landlord's estate may have passed to the Crown in this way, enquiry should be made of the Treasury Solicitor. *Bona vacantia* will also apply to the property of a person who dies intestate and with no known next of kin.

The task of tracing a named individual who may be living anywhere in the world, if they are alive at all, is that much more difficult. If the landlord is unregistered, leaseholders will have to begin their search from the last known ground rent or service charge receipts — following up all areas of enquiry. But however impossible the task, it is essential that leaseholders keep a meticulous record of all enquiries made, and the outcome of those enquires, so that this can be proved to the relevant court or tribunal.

Dealing with Landlord Default

When tenants default on their leasehold obligations, landlords have a choice of remedies.

Provided ground rent or service charges have been correctly demanded, landlords can sue immediately for any sums which are unpaid. Once judgment is obtained for the arrears, the landlord will have the full range of county court remedies, including use of bailiffs or earnings attachment. Other options include bankruptcy for larger arrears and injunctions to deal with non-monetary lease infringements. This is not to forget forfeiture, which remains the ultimate weapon in a landlord's armoury. But what is a leaseholder to do when it is the landlord who is in default of his obligations relating to the management, insurance or repair of a property?

Withholding rent

Can a leaseholder lawfully withhold payment of rent (or service charges) in such circumstances? Technically the answer is "no". A tenant's covenant to pay rent and a landlord's covenant to repair are independent. A breach of one does not excuse a breach of the other. Withholding rent may also put a leaseholder in a worse position by triggering a landlord's right to take default action. Many leases also stipulate that rent must be paid "without deduction". But note that "without deduction" was held in *Connaught Restaurants Ltd* v *Indoor Leisure Ltd* [1993] 2 EGLR 108 *not* to exclude a tenant's right to set off financial liabilities owed to him by the landlord unless such right of set off is also specifically excluded. There are only two circumstances when damages for breach of a landlord's covenant to repair can be set off against a future liability to pay rent.

After considering a long line of cases, Goff J held in *Lee-Parker* v *Izzet* [1971] 3 All ER 1099 that a *legal* right of set-off arose when a tenant had already spent money on repair which were the landlord's responsibility, so long as the tenant had previously alerted the landlord to the need of those repairs and the amount spent was fully quantified and could not be challenged.

In *British Anzani (Felixstowe) Ltd* v *International Marine Management (UK) Ltd* [1979] 1 EGLR 65, Forbes J extended this principle by stating that an *equitable* right of set-off could still arise even when a tenant had not actually spent money carrying out repairs. All that had to be proved was that the tenant had a genuine cross claim for unliquidated damages and that there was no other legal remedy. But a right to set-off can only apply where both the claim for rent and the counterclaim for damages are linked by the same landlord and tenant relationship. In *Anzani*'s case a collateral guarantee given by the landlord to rectify defects in the floors of two warehouses were so closely connected with the leases that it would have been manifestly unjust not to allow set-off.

Risk assessment

Even where leaseholders believe that they have the right to set-off claims for damages against future rent liability, it is essential that a full risk assessment is first carried out. Before withholding rent it is always prudent to notify the landlord of the leaseholder's intention to withhold payment and why it is being withheld. Unless the leaseholder has already spent money in carrying out repairs which were the landlord's responsibility, any rent withheld should be set aside so that it can be paid to the landlord immediately the problem is resolved. In situations where ground landlords have completely washed their hands of their obligations it may be better for those leaseholders to formally take over management under their new "right to manage" (see Chapter 6).

Correct procedures

The need for leaseholders to follow the correct statutory procedures before attempting to usurp a landlord's rights and responsibilities was underlined by Etherton J in *Metropolitan Properties Co Ltd* v *Wilson* [2002] EWHC 1853, Ch.

The leaseholders of 10 Hyde Park Mansions, St Marylebone, had long complained about the state of repair of their building and it was not disputed that the exterior and interior were in need of repair. Service charges had been withheld and Metropolitan Properties were now suing the lessees for those arrears and for forfeiture of their leases. The leaseholders also intended to carry out the repairs themselves and had commissioned their own building contractor to erect scaffolding on the exterior of the building. The existence of this scaffolding now prevented Metropolitan from having access to the building for its own maintenance and repair and they sought an injunction requiring the leaseholders to remove it.

Metropolitan had put forward a scheme of repairs for which it had obtained quotations. Having complied with the statutory procedures for consulting leaseholders (see Chapter 5), Metropolitan instructed its own contractor to carry out the external works.

Etherton J granted the injunction sought by Metropolitan. He said that the existence of the scaffolding was a trespass as the exterior of the building and communal areas were not included in any of the residential leases (as explained in Chapter 1). Neither had the leaseholders notified Metropolitan of their intention to erect the scaffolding if the landlord did not commence the same repairs in the near future. To permit the leaseholders to carry out their own repairs to the exterior of the building would mean permitting work for which there had been neither consultation nor agreement with the landlord as to the selection of the proposed contractor or the terms of the contract. As the cost of the leaseholders' work would inevitable be deducted from their service charges, this would alter the contractual leasehold framework which makes the landlord responsible for such repairs and also supplant the statutory procedures enabling leaseholders to challenge the reasonableness of service charges in the light of the quality of the work carried out.

Specific performance

Until Pennycuick V-C's decision in *Jeune* v *Queens Cross Properties Ltd* [1973] 3 All ER 97 it was thought that no judge would order specific performance to compel a landlord to comply with its repairing obligations. But Pennycuick saw no reason why specific performance could not be ordered to force a landlord to reinstate a york stone balcony, as such an order could be framed in precise terms telling the landlord exactly what it had to do to comply.

Damages

The basic common law sanction for a landlord's failure to comply with its contractual obligations is an action for damages. All that generally has to be proved is that the landlord has failed to comply with significantly with its obligations under the lease and that a leaseholder has suffered loss as a result. Such loss may be purely financial or involve personal injury or interference with the peace and comfort which the leaseholder is contractually entitled to expect. Where the claim is based on a failure to repair, it may also have to be proved that the leaseholder took adequate steps to make the landlord aware of the defect.

Financial loss might include any depreciation in the value of the leasehold interest resulting from the landlord's failure to comply with its contractual obligations. For small but important items of disrepair, it might be the cost to the leaseholders of putting it right themselves — if this is a reasonable course of action for the leaseholders to take in the circumstances and the landlord is kept informed.

Sale, Sublettings, Alterations and Use

Most long residential ground leases are freely assignable.

This means that the lease can be sold (as a whole) on the open market without the need for the landlord's permission or any other restrictions. This makes good sense. As it is the lessee who will own almost all the equity (capital value) of the residential unit, there is no reason why a ground landlord should expect the right to control how and by whom the unit is occupied. But there are exceptions.

For a minority of residential leases, landlord's consent will be required before the unit can be sold. This might apply to special types of accommodation, such as retirement housing or a shared ownership. Even if there are no restrictions on the sale of the unit, there may be restrictions on sublettings.

It is almost certain that any residential lease will contain restrictions on the right of a leaseholder to carry out alterations or to engage in non-residential use. Each of these areas of restriction is now considered in more detail.

Absolute or qualified

Restrictions on sale, subletting, alterations or change of use can be either absolute or qualified.

An obligation on the part of a leaseholder, "Not to allow the Premises to be occupied by any person under the age of 55", is absolute. That is to say, it is non-negotiable. The tenant does not have the right to even ask the landlord for permission to allow occupation by a younger person.

A qualified covenant might read "Not *without the landlord's prior written consent* to allow the Premises to be occupied by anyone under the age of 55."

The second example does not completely bar occupation by a younger person. It means that before allowing such occupation, the leaseholder must first seek the landlord's prior written permission. Where an obligation is qualified in this way the Landlord and Tenant Act 1927 will imply a term that a landlord's consent to a tenant's request cannot be unreasonably withheld: see below.

Restrictions on assignment/underletting

Any well-drawn lease will absolutely prohibit assignment (sale) of *part* of the residential unit. If this were not the case, it would be possible for a lease to be *split*, which could in turn create complications as regards the management of the block. Therefore partial assignments should always be prohibited. This prohibition may or may not extend to partial sublettings depending on the terms of the lease.

Where there is a qualified covenant against the assignment or subletting of the whole of the premises without the landlord's consent, section 19 Landlord and Tenant Act 1927 states that such requirement shall be subject to a proviso: "that such consent is not to be unreasonably withheld". This does not preclude the landlord's right to require reimbursement of any legal or other expenses incurred by the landlord in connection with any application for such consent.

But the 1927 Act did not provide guidance as to what constituted an unreasonable withholding of consent. Such guidance came more than half a century later in the case of *International Drilling Fluids Ltd* v *Louisville Investments (Uxbridge) Ltd* [1986] 1 EGLR 39 which set out the following principles which any court had to apply.

1. The purpose of a covenant against assignment [or underletting] without landlord's consent is to protect the lessor from having his premises used or occupied in an undesirable way or by an undesirable tenant or assignee.

2. It follows from the above that a landlord is not entitled to refuse consent to an assignment on grounds which have nothing whatsoever to do with the relationship of landlord and tenant as regards the lease. Landlord's consent cannot be withheld to achieve a collateral purpose unconnected with the terms of the lease.

3.   The onus of proving that consent has been unreasonably withheld is on the tenant; [reversed by the Landlord and Tenant Act 1988 in relation to assignment and underletting].

4.   It is not necessary for the landlord to prove that the conclusions which led him to refuse consent were justified, if they were conclusions which might have been reached by reasonable landlord in the circumstances; [reversed as above].

5.   It may be reasonable for a landlord to refuse consent to an assignment because of the way the proposed assignee intends to use the premises, even though that proposed use may not be forbidden by the lease.

6.   A refusal of consent might be unreasonable if the effect of that refusal would be disproportionately damaging to the tenant, compared with how a grant of consent would have affected the landlord's interests.

Although *International Drilling Fluids Ltd* provided legal clarity, a leaseholder seeking consent to assign (or underlet) still faced the following difficulties:

*   it was not necessary for the landlord to prove that his reasons for refusing consent were "reasonable" unless the landlord's position was so unreasonable that no other *reasonable* landlord could have behaved in a similar way
*   it was for the tenant to prove unreasonableness — not vice versa
*   there was no obligation on the landlord to respond promptly to tenants requests for consent.

In practice a leaseholder seeking landlord's consent is more likely to be faced with delay on the part of a landlord than a prompt refusal. It is in consequence of such delays, and the uncertainties which go with them, that sales may be lost. No prospective purchaser will wait forever to see whether they are approved by the landlord.

This unsatisfactory situation was rectified by the Landlord and Tenant Act 1988, which applies to any tenancy containing a covenant on the part of the tenant not to assign, underlet, charge (meaning to mortgage) or part with possession of the premises in any other way without the landlord's consent. The Act sets out a formal procedure which landlords must follow when receiving a tenant's application for their consent to a proposed assignment, underletting, charging or parting with possession of premises.

Section 1(3) of the Act states (in summary) that a landlord receiving a tenant's written request for landlord's consent must, *within a reasonable time*:

a.   give consent, unless there are reasonable grounds for refusing consent and
b.   give the tenant written notification of his decision (whether granting or refusing consent) and specifying any conditions to which the consent is subject — or if consent is withheld, the reasons why it has been withheld.

Section 1(6) imposes the following additional duties on the landlord in relation to a tenant's request for consent (in summary):

a.   to demonstrate (if required) that any consent was given within a reasonable time
b.   to demonstrate that any condition imposed on the consent was reasonable and
c.   if consent was refused — to show that the refusal was reasonable, if required to do so.

In some cases, a landlord may himself be a lessee under a "superior lease" and may be unable to grant consent on the tenant's request without himself being able to obtain consent from the superior lessor. In such cases section 2 of the 1988 Act requires the landlord to pass on the lessee's application to the superior landlord — who will then be under similar duties as to how the tenant's application is to be dealt with.

Section 4 states that any non-compliance by a landlord (or superior landlord) with his responsibilities under the Act may be made the subject of civil proceedings for "breach of statutory duty". This means that the tenant can apply to a county court for a "declaration' that landlord's consent has been unreasonably withheld — or unreasonably delayed — in either of the following circumstances:

•   if the landlord has failed to respond or make any formal decision on the application within a "reasonable time";or
•   if consent has been refused (or been granted subject to unacceptable conditions) and either the landlord has failed to state the reasons for such refusal or that the grounds for such refusal (if stated) or the conditions imposed, do not appear reasonable.

As well as a "declaration", section 4 also opens up the possibility of a claim for damages where a leaseholder has lost out financially as a direct result of any wrongful refusal or delay on the part of the landlord in giving consent.

What the 1988 Act does not do is to stipulate the exact time-scale within which a landlord must respond. This will be an issue for interpretation by the courts having regard to the circumstances in each individual case. Neither does the Act provide further guidance on the grounds on which landlords can validly withhold consent. *International Drilling Fluids* therefore remains the principle reference point for such issues.

How the 1988 Act helps tenants is by setting out the procedure which landlords must follow and by giving tenants the information they need to challenge any decision which they perceive to be unreasonable. The other major reform brought about by the 1988 Act is that it is no longer for the tenant to prove "unreasonableness" on the part of the landlord. Instead landlords must prove that their own decisions are "reasonable".

Consent might be validly withheld if the landlord was not reasonably satisfied about the ability of a prospective assignee to meet the financial obligations involved in taking on the particular lease. Obvious reasons for which consent could not normally be withheld are on grounds of a prospective assignee's (or sublessees) race, colour, religion, sex, sexual orientation, marital status, disability, etc (see below). But there are also many grey areas.

Is it reasonable for a landlord to refuse consent for an assignment on the grounds that the lessee has failed to comply with his obligation to keep the property in decorative repair? It could be argued that a landlord might be glad to get a new tenant more ready and willing to comply with outstanding covenants. However in the case of *Orlando Investments Ltd* v *Grosvenor Estates Belgravia*, [19889] 2 EGLR 74, the Court of Appeal held that it was not unreasonable for a landlord to refuse consent for an assignment when he was not satisfied that the prospective assignee was in a position to put the necessary repairs in hand and the assignee was unwilling to accept the landlord's condition requiring a performance bond or deposit.

A landlord who is minded to grant consent to a proposed assignment (or under letting) may first of all write back to the leaseholder stating that it has no objection 'in principle' to the proposed transaction subject to the prospective assignee or underlessee entering into a formal licence to assign or licence to underlet, whereby the assignee or underlessee enters into a direct contractual arrangement with the landlord to observe and perform the tenant's (or undertenant's) obligations contained in the lease (or underlease) with effect from the date of the assignment or underletting.

On completion of the assignment (or underletting) or any other transaction affecting the leasehold interest (whether landlord's consent is required or not), the assignee (or lessee in the case of an underletting) will normally be required to notify the landlord of completion of the transaction, giving required particulars, and paying a nominal registration fee to cover the landlord's costs in recording those details.

Original tenant liability

Many former holders of leases executed before 1996 will not be aware that they can still be sued under the terms of that lease, even if the default occurs long after they have transferred their interest to someone else.

This legal concept, known as original tenant liability arises out of the privity of contract which arose between the original landlord and the original lessee when the lease was first signed. It reflects the fact that any lease is primarily a contract between two parties which is designed to subsist over the full term of the lease. Therefore the fact that the tenant chooses to sell the lease after a few years may not automatically release him from liability so far as the landlord is concerned. The same principles apply if it is the landlord that sells his freehold (or superior leasehold interest) during the currency of the lease. That landlord will technically remain answerable to the same original tenant during the whole of the term of the lease for any future breach of the landlord's contractual obligations.

In practice, a long residential ground lease may be sold on many times during its currency — as may the landlord's freehold or superior leasehold interest. Therefore this ancient rule of law made little practical sense. It could also operate unfairly by making former tenants (and landlords) liable for future default over which they have no control.

This unfair liability was modified by the Landlord and Tenant (Covenants) Act 1995 — but only in relation to new leases granted after that Act took effect on 1 January 1996. These are known throughout the Act as new tenancies.

Section 5 of the 1995 Act states (in summary) that where a new tenancy is assigned, the outgoing tenant is automatically released from his future obligations under the tenancy. However, the same does not apply to former landlords. Any landlord seeking release from future liability under a new lease must follow the procedure set out in section 8 of the Act, which requires the landlord to serve notice on the tenant formally requesting release. Such notification must be given either before or within four weeks of completion of the transfer of the landlord's interest. The tenant then has four weeks to object to such release — failing which the release will take effect automatically. If the tenant does object (and that objection is not withdrawn) such release will be dependent on the landlord (or former landlord) obtaining a declaration from the county court that it is reasonable for the landlord's future liability to be released.

For leases signed before1996 ("old leases"), privity of contract will remain and original landlords and original tenants will face continued liability long after they have sold on their respective interests. But even former tenants under old leases will be assisted by section 17 of the Act, which will exonerate them from liability to make good any future arrears of rent or service charge unless (in respect of those arrears) the landlord has, within six months of those arrears falling due, informed the former tenant of those arrears and the amount which the landlord intends to collect from that former tenant. But note that section 17 only relates to financial default, not to other breaches for which a former tenant could be held liable, (such as disrepair).The 1995 Act also introduced two other features:

- section 16 — authorised guarantee agreements — whereby as a condition of obtaining landlord's consent to an assignment, the former tenant (under a new lease) agrees to make good any future default by their immediate assignee (but not after any further assignment) and
- section 19 — overriding leases — whereby any former tenant who is required to make good any arrears of an assignee (or any other future owner of the lease) is entitled to require the landlord to grant him an overriding lease — in effect making the defaulting assignee into a subtenant of the former tenant. An overriding lease is an intermediate lease standing between the existing landlord and the occupational lease. Its effect is to put the former tenant back into control and enables him to evict the assignee and take back possession of the premises if default continues or reoccurs.

In practice original tenant liability has always been more of a problem for former commercial tenants than former residential lessees. This is because the capital value of a residential ground lease provides its own security for performance of a leaseholder's financial and other obligations. Faced with persistent default it would make more sense for a residential ground landlord to go for forfeiture rather than waste time trying to chase former tenants. It is perhaps no coincidence that there are no recent reported cases on original tenant liability within the setting of a residential ground-lease.

Alterations to premises

Almost all leases restrict in some way the right of leaseholders to carry out alterations to their premises. Such restrictions are necessary to protect the stability of the building within which the residential units are situated, to avoid shoddy workmanship and also to ensure that any alterations are in keeping with the general style of the building.

As already explained in Chapter 1, some parts of a residential unit, such as the external walls, internal load bearing walls, and floor joists, may not even belong to the lessee. Therefore the option of carrying out alteration to those items does not arise. A well drawn lease will, in any event, prohibit absolutely any alterations to anything which may affect the structural stability to the building. Other alterations to individual residential units may be permitted provided the written consent of the landlord is first obtained. As with qualified restrictions on assignment, where landlord's consent is required for proposed alterations, that consent cannot be unreasonably withheld.

Section 19 (2) of the Landlord and Tenant Act 1927 (in summary) states that where a lease contains a covenant, condition or agreement against the making of *improvements* without landlords consent, there is a statutory proviso that such consent is not to be unreasonably withheld.

The principles outlined in *International Drilling Fluids Ltd* v *Louisville Investments (Uxbridge) Ltd* [1986] (see above) will apply to applications for consent to alterations in the same way as they apply to applications for consent to assign or underlet. But the leaseholder applying for consent to alterations will not benefit from the procedural reforms introduced by the Landlord and Tenant Act 1988, as those reforms only apply where consent is sought for a proposed assignment, mortgage or underletting. Therefore the leaseholder seeking consent for alterations will be in a worse position, as the landlord will be under no obligation to respond promptly or to prove that he had acted reasonably in refusing consent. However in the High Court case of *Luzatto* v *Danzig*, Lawtel 19 June 2002, Judge John Jarvis QC ruled that a landlord had acted unreasonably in refusing consent to a leaseholder's proposed roof extension.

Sandra Luzatto's 99 year lease of 12 Marylebone Mews, London W1, contained a clause that the tenant, 'shall not at any time during the said term without the licence in writing of the lessor first obtained, make any alteration in the plan or elevation of the demised premises or in the party walls or in the principal or load-bearing walls or timbers thereof'. By section 19(2) of the 1927 Act, there was implied proviso that such consent should not be unreasonably withheld.

Mrs Luzatto wished to add a second floor roof extension, make alterations to front and rear elevations, and install a new skylight to the rear flat roof. Its effect was to add a new floor to provide additional bedroom accommodation and two bathrooms — thus enabling the first floor to be utilised for other living accommodation. Planning permission for the alterations had previously been obtained from the City of Westminster.

The landlord, Mr Danzig, objected on the grounds that the new windows would overlook another property belonging to him at 66 Wimpol Street, which was divided into six flats. A second objection was that the raising of the roof of Mrs Luzatto's property would lead to a loss of light to the rear windows of 66 Wimpol Street. There was a third issue relating to the hot water supply, which was resolved by agreement before the trial. The issue relating to the windows was also resolved by making them frosted, with an opening of only 100mm. The only remaining issue, which the court had to determine, was whether the anticipated loss of light provided reasonable grounds for the landlord to object.

It was apparent from undisputed expert evidence that there would be no loss of sunlight as the building behind the premises, 12 Welbeck Street, was already substantially higher than the premises, so that the increase in height would cause not direct sunlight loss. However ambient light loss would be 0.7% or 2% of the existing light. The judge also considered that the drawings and plans of the proposed alterations were ascetically attractive and would enhance the property and surrounding area.

Having reviewed the *International Drilling* case and other previous judicial decisions, Judge Jarvis found any loss of light to be minimal in the extreme and that such loss would imperceptible to the human eye.

Neither could it be suggested that the alterations would in any way reduce the value of the landlord's interests. As such he ruled that it would be wholly unreasonable for a landlord to refuse consent in those circumstances. He also said it would be wholly out of proportion to take into account such a tiny loss of light against the advantage to the tenant in making the alterations.

Assuming that a ground landlord is willing to grant consent for alterations, there remains the question of how that consent should be documented. For the simplest alterations, consent might be given in the form of a letter. For more substantial alterations, it would be in the mutual interest of the landlord and tenant for such consent to be recorded in a licence for alterations to which will be affixed detailed plans and specifications of the proposed alterations. A standard condition of any landlord's consent will be that the tenant complies with all relevant town planning and building regulation requirements, both before and in the carrying out of the work. In whatever manner the landlords consent is documented, it is important that evidence of such consent is kept with the title deeds in case of future query. This applies both to the landlord and the tenant.

Often the first hint that unauthorised alterations may have been carried out is when a lease is in the process of being sold. Lawyers acting for the prospective purchaser will, as part of their standard pre-contract enquiries, ask whether there have been any alterations to the premises and — if so — whether landlords consent has been obtained in accordance with the provisions of the lease. If the answer is "yes" — "there had been alterations" - but "no" — "the landlords consent has not been sought or obtained" — the question is what to do next? Applying for landlord's consent retrospectively for previous alterations raises the risk that consent may be refused and the tenant being required to reinstate the premises to their original condition. It is therefore an issue on which professional advice should be sought. Another solution maybe to seek title indemnity insurance provided that this is acceptable to the relevant lending institution.

Alterations to accommodate disability

Special provisions apply to improvements required by disabled residential occupiers to facilitate their enjoyment of their home.

In those circumstances section 49G of the Disability Discrimination Act 1995 (as amended by the Disability Discrimination Act 2005) prevents the landlord from unreasonably withholding or delaying consent to such alterations. However, this only applies where a lease allows the leaseholder to make alterations approved by the landlord and provided the leaseholder has applied in writing for such consent.

A landlord who refuses consent must give the leaseholder a written statement of the reason why consent has been withheld. If an issue arises as to whether a landlord was acting reasonably either in refusing consent or as regards any condition imposed on any consent, it is for the landlord to prove that he had acted reasonably. If consent is unreasonably withheld the leaseholder is entitle to proceed as if consent had been given. If consent is granted subject to an unreasonable condition, the leaseholder may treat the landlord's decision as a refusal.

Change of use

It is rare that a residential leaseholder would seek permission to change the use of the unit from residential to business use. But this situation might arise where a leaseholder wishes to use part of the living accommodation for office purposes or as a consulting room.

Usually any such change of use would be absolutely prohibited within the terms of the lease. However if such use is permissible with landlord's consent, section 19(3) of the 1927 Act states that the ground landlord cannot charge any money for granting such consent beyond reasonable legal and other professional expenses incurred by him in dealing with such requests. However, unlike assignment and alterations, there is no statutory proviso that consent should not be unreasonably withheld.

Anti-discrimination provisions

Section 31(1) Sex Discrimination Act 1975 makes it unlawful for a landlord to discriminate against a man or woman by withholding consent for a proposed lease assignment or underletting on grounds of a person's gender.

Section 24 of the Race Relations Act 1976 contains a similar prohibition against discrimination by a landlord on grounds of race, ethnic or national origin.

Section 22(4) of the Disability Discrimination Act 1995 makes it unlawful for a landlord to discriminate against a disabled person by withholding his license or consent for the disposal of the premises to the disabled person.

Checklist for applying for landlord's consent

1. *Check the terms of the lease.* Is landlord's consent actually required? Does the lease set out any criteria relating to the grant of such consent? Is any covenant "absolute" or "qualified" (see above)?

2. *Anticipate what information the landlord will reasonably require to deal with the request.* Where the application relates to a proposed assignment or subletting a landlord will require satisfactory references for the prospective assignee or undertenant. For alterations, provide adequate scale drawings and specifications. If in doubt as to likely requirements, clarify this informally with the landlord's agent before making the request. As a matter of courtesy any written application for landlord's consent will also contain an undertaking by the tenant to reimburse the landlord's *reasonable* professional fees in considering the request and in documenting any consent.

3. *Assess a reasonable time within which to expect a landlord's response.* This is important as under the Landlord and Tenant Act 1988 the landlord is required to respond within a reasonable time to applications for consent to assign, underlet etc.

4. *Make a formal written application for consent.* This should refer to the relevant clause in the lease and also the Landlord and Tenant Act 1988 (where applicable). The application should also be accompanied by everything the landlord will need to determine the application (e.g. references, for prospective assignee/under-lessee; plans and specifications of proposed alterations etc).

5. *Ensure that the formal application is promptly acknowledged by the landlord or (landlords agent) or that delivery of the application to the landlord and its enclosures can be proved* (eg by use of recorded delivery).

6. *Monitor progress on the application.* Send reminders if necessary.

7. *If landlord (or landlord's agent) does not respond at all within a reasonable time, take legal advice.* The same applies if there is subsequent unreasonable delay on the landlord's part in making a decision. Ensure that any reasonable requests from a landlord for further information are promptly complied with.

8. *Consider the landlord's substantive response.* If the landlord's response is unfavourable, analyse the landlord's decision, and in particular the reasons given. Is there scope for negotiation? Take legal advice.

9. *If the landlord expresses his willingness to grant consent in accordance with the application, arrange for this to be documented.* Save where this can be dealt with by letter, a formal licence to assign/underlet, or licence for alterations may be required to evidence the fact that consent has been given and the terms on which it has been given. This is a standard document which is normally prepared by the landlord's agent or lawyer — but at the cost of the leaseholder who is applying for such consent. See also Chapter 14 which sets out the procedure by which a tenant can challenge any legal or other administrative charges which are perceived to be unreasonably high.

10. *Place the original licence with the deeds.* And for future reference, keep a copy to hand.

Service Charges

Service charges are one of the biggest areas of dispute between lease-holders and ground landlords — as well as between leaseholders and other leaseholders. The amounts involved can be sudden and substantial and take leaseholders by surprise. It is also something over which many leaseholders feel that they have little control and are left to pick up the bill.

However since 1985 residential leaseholders have had the statutory right to be consulted on, (though not veto) proposed service charges. Since 1996 it has been made easier for leaseholders to challenge service charges which they believe to be excessive and unreasonable, without expensive litigation. Statute now prevents a ground landlord from threatening forfeiture for unpaid service charges unless the amount involved has been agreed or adjudicated upon. Leaseholders also have a statutory right to receive authenticated service charge accounts and to inspect receipts, invoices and other service charge items. But the starting point remains the lease itself.

Contractual provisions

A modern residential lease is likely to contain a schedule setting out, at length, those items relating to the management, maintenance, repair and insurance of the building, for which the ground landlord is entitled to claim reimbursement from each leaseholder. Conventionally such schedules are widely drawn to include finance charges, professional fees and administration costs — to ensure that the landlord can never be out of pocket. Schedules may also include a "sweeper clause" entitling the landlord to add to or vary the listed services.

The leases should also set out the proportion to which each individual leaseholder is required to contribute towards the landlord's costs in providing services to the building as a whole. Such proportions are usually expressed as a percentage. These may either be equally split between the number of leases or (where the flats differ in size), fixed according to the floor area of each flat. However individual proportions are fixed, added together they should equate to 100% of the landlord's costs.

Finally the lease will contain a mechanism for calculating the service charge payable by each leaseholder and for resolving disputes. The more flats comprised within the building or on the estate, the more complex the procedures are likely to be. Most service charge revisions will require something to be paid on account towards anticipated service charge costs, with a final settling-up at the end of the service charge year. Any balance maybe either carried forward or an additional demand made when the final figures are known. The mechanism may also require delivery of a professionally authenticated service charge statement annually to each leaseholder. There may also be provisions for resolving any dispute over service charges by arbitration. But note that any dispute — resolution provisions contained in the lease will no longer override the statutory rights which residential leaseholders now have to challenge the reasonableness and lawfulness of service charges through an LVT (see below).

Although service charge provisions may be drafted on a "catch-all" basis, the courts and LVTs will interpret them restrictively. That is to say that any ambiguity in the service charge provisions will be resolved in the leaseholder's favour. It means that a landlord will not be able to recharge for a disallowed item even if he has incurred expenditure providing that service. The Court of Appeal decision in *Gilje* v *Charlesgrove Securities Ltd* [2002] 1 EGLR 41 explains the point.

The leaseholders owned five flats within 27 Lennox Gardens, London SW1. The sixth flat was in the basement and occupied by the caretaker. Each of the leases included an obligation on the lessor to provide a resident housekeeper or porter to be in attendance between 7am and 12 noon and at other times for the proper fulfilment of specified duties. The costs to the lessor for providing the caretaker was a service charge component.

The issue for the court was whether the lessor could also recover from the leaseholders the notional rental value of the caretaker's flat, which was owned by the lessor and provided to the caretaker free of charge as part of her employment package. The Court of Appeal ruled that the lessor could not recharge those notional costs — even if the caretaker's wages were artificially increased to include such notional rent, which was then deducted via a book entry.

The judges noted that the service charge provisions made no reference to the value of the caretaker's accommodation as being out of pocket expenditure on the part of the landlord. Laws LJ said that the real question was whether the notional rent on the caretaker's flat fell within the expression "monies expended" in the service charge provisions. However, there had to be clear contractual provisions entitling the lessor to do this. A reasonable leaseholder reading the document would not perceive that the service charge provisions obliged him to contribute to the notional cost to the landlord of providing the caretaker's flat. There are two other cases, which demonstrate the court's approach.

- *Pole Properties Ltd* v *Feinberg* [1981] 2 EGLR 38 — when the Court of Appeal held that where events occurred for which the parties had made no provision and which were outside the realm of their speculation, the court must apply what was fair and reasonable as regards service charges. In that case there was an argument over the amount of a leaseholder's obligation to contribute towards an increase in the heating costs of a block of flats (based on a floor area), in circumstances where there had been the addition of another building to the original block. It was held that payment should be regulated according to the tenant's use of the heating, not floor space.
- *Boldmark Ltd* v *Cohen* [1985] 1 EGLR 47 — when the Court of Appeal prevented lessors from recovering interest payments on borrowing for "general administration and management" of a block of flats. Although the court saw no reason, in principle, why lessors should not recover such expenditure, the service charge provisions in the leases were not wide enough to cover the recovery of interest for notional borrowing in circumstances where the lessor had funds to pay for general administration and management. The landlords had failed to prove that the leaseholders had contracted to pay such interest.

Consultation requirements

A lessor who does not follow statutory consultation procedures (see below) in relation to significant or long-term service charge items will be unable to recover those costs (above a minimum level) unless such recovery has been sanctioned as "reasonable" by an LVT.

The law is set out in section 20 of the Landlord and Tenant Act 1985 (as rewritten by section 151 CLRA). The amended section 20 states (in summary) that the cost of specific and long term service charge items, above a statutory limit, will be irrecoverable from leaseholders unless either the lessor has complied with the statutory consultation requirements or that an LVT has agreed to dispense with such requirements. The consultation requirements themselves are set out in the Service Charges (Consultation Requirements) (England) Regulations 2003. Relevant provisions are summarised below.

- Regulation 4 — limits any leaseholder's annual contribution towards a qualifying long term agreement to £100 unless the consultation requirements have been complied with. A qualifying long term agreement is one entered into by a lessor with a contractor for a period of more than 12 months.
- Regulation 6 limits to £250, each individual leaseholder's requirement to contribute to towards other service charge items unless there has been strict compliance with the consultation requirement.

Qualifying long term agreements

Schedule 1 to the regulations sets out the consultation requirements for a qualifying long term agreement. These require the lessor to give written notice of his intention to enter into the agreement to each leaseholder and (where it exists) to a recognised tenants association. This notice must generally describe the goods or services to be provided or the work to be carried out under the proposed agreement or specify a place and time at which that information may be inspected. The notice must also state the lessor's reasons for considering it necessary to enter into the agreement and (where relevant) the lessor's reasons for considering it necessary to carry out specified works. The notice must also invite the leaseholders' observations in relation to the proposed agreement and specify the relevant consultation period and the address to which those observations must be sent. The notice must also invite each leaseholder and recognised tenant's association (if any) to propose the name of someone from whom the lessor should also try to obtain an estimate in respect of the relevant matters.

Where a single nomination is made either by a tenants' association or by the leaseholders, the lessor must try to obtain an estimate from that person. Where leaseholders make more than one nomination, the lessor must try to obtain an estimate from at least one of those nominees.

From those estimates the lessor must prepare at least two proposals in respect of the relevant matters — including at least one proposal based on an estimate from someone wholly unconnected with the lessor. Each proposal must contain a statement of the proposed contractor's name and address and detail any other connection between that proposed contractor and the lessor. Such a connection will be presumed where the lessor is itself a company and the proposed contractor is another person or company closely associated with the lessor.

If it is reasonably practicable to do so, the lessor must also estimate the relevant contribution which would be attributable to each leaseholder. Where it is not reasonably practicable for the landlord to specify the exact financial contribution required from each leaseholder, it might still be practicable for the landlord to ascertain and quote the relevant unit cost or hourly or daily rate applicable to the relevant matters. Notification of the proposal must also provide details of any agent appointed by the lessor to discharge any of the lessor's obligations to the leaseholders as regards the management of the premises to which the agreement relates — including whether the person belongs to a professional body or trade association or subscribes to any code of practice or government accreditation scheme. The proposal must also state the intended duration of the proposed agreement.

Where observations are received from leaseholders or a recognised tenants association, the landlord is required to include a statement summarising those observations and the lessor's response to them. A second round of consultation will then take place.

The landlord will be required to give written notice of his proposals to each leaseholder and tenants' association (where applicable), accompanied by either a copy of the proposal or specifying a time and place at which they may be inspected. As before, the notice must also provide the recipient with an opportunity to make observations.

Within 21 days from signing the contract, the lessor must again write to each leaseholder and tenants' association (if any), stating his reasons for entering into that contract or specifying a time and place at which a statement of those reasons may be inspected. If observations were previously received from leaseholders or an association, the lessor must also summarise those observations and his response to them or specify time and place for inspection, as before.

Public sector landlord

Schedule 2 applies to the situation where a public sector landlord is required to publish notice of a proposed contract in the *European Journal*, in compliance with European Procurement Regulations. Under European law, a public sector organisation intending to commission works, supplies or services above a prescribed value must tender those works, supplies, or services on a Europe wide basis. Therefore any leaseholder-consultation must also be compliant with European law.

Schedule 2 repeats the requirements for lessors to give notice of their intention to enter into qualifying long term agreements to each lease-holder and tenant's association (where applicable). But neither leaseholders nor tenants associations will be invited to nominate a particular contractor, as this would be incompatible with European law. In other respects the procedure is similar.

Schedule 3 applies to the situation where qualifying works are to be carried out pursuant to an existing qualifying long term agreement. As before, preliminary notice of intention must be given to each leaseholder and tenant's association (where applicable), giving reasons and inviting observations.

Having had regard to such observations in relation to the proposed works and estimated expenditure, those works can then be commissioned. Within 21 days from such commissioning the lessor must give notice to the persons making those observations together with his response.

Stand alone works

Schedule 4 deals with the situation where stand alone qualifying works are proposed. Part 1 deals with the public sector scenario where notice has to be published in the *European Journal* (see above). Under part 1 the lessor must give preliminary notice of his intention to each leaseholder and tenant association describing the proposed works or specifying a time and place at which the information may be inspected and stating the landlord's reasons for carrying out the proposed works. Although leaseholders will be invited to make general observations in relation to the proposed works, they will not be entitled to nominate anyone from whom an estimate is to be obtained.

Having had regard to any observations received, the lessor must then prepare a Contract Statement of Proposed Works giving details of the person with whom the lessor proposes to contract and details of any other connection between that contractor and the lessor, as well as giving appropriate financial estimates. The lessor must then give written notice of his intention to enter into the proposed contract to each tenant and tenant association accompanied by his contract statement and inviting observations.

Where observations are received pursuant to the above notice, the lessor must within 21 days of such receipt, notify the person making those observations of his response.

Part 2 of Schedule 4 deals with the most common situation where European procurement regulations do not apply and the proposed works are not pursuant to a qualifying long term agreement. As in all other cases, the lessor must give preliminary notice of his intention to each leaseholder and tenants' association describing the proposed works in general terms, stating why it is necessary carry out the proposed works; and inviting receipt of observations from leaseholders and tenants' associations. In this situation, the notice must also invite each leaseholder or tenant's association to propose a contractor of their own from whom they would wish the lessor to obtain an estimate.

Having had regard to any observations received, the lessor must try to obtain an estimate from any lessees' nominee, or where there is more than one nominee, from the person who received the most nominations. The lessor must then supply a statement, in respect of at least two of the estimates, setting out the amounts specified in those estimates as well as a summary of observations received and his response to those observations. At least one of those estimates must be from someone wholly unconnected with the lessor. That statement must be copied to each leaseholder and tenants association providing an opportunity for inspection of those estimates and receipt of observations. Having had regard to those observations and having signed the contract, the lessor must within 21 days give written notice to each leaseholder and the tenant's association (if any) stating his reasons for awarding the contract or specifying a time and place where those reasons may be inspected and — where observations have been received — summarising those observations and his response to them.

Insurance

Section 30A of the schedule to the 1985 Act apply where a service charge includes a component for buildings insurance.

Para 2 of the schedule then entitles any leaseholder to insist on a written summary of the insurance cover. On receipt of a written notice requiring such information, the landlord must provide within 21 days a summary which includes: the amount insured; the name of the insurer; and the risks covered.

Para 3 the entitles the leaseholder to insist on reasonable facilities to inspect and take copies or make extracts of any policy. Where the insurance was taken out by a superior landlord, an immediate landlord receiving a leaseholder's request must take steps to obtain the insurance information from the superior landlord.

Para 6 makes it an offence for a landlord to fail, without reasonable excuse, to respond to a requirement under this schedule.

Where insured damage has been caused either to an individual residential unit or to the building within which it is situated, para 7 allows the leaseholder to contact the insurer directly about the damage.

In circumstances where a lease obliges a tenant to insure with an insurer nominated by the landlord, para 8 allows the tenant to apply to the county court or LVT to challenge the landlord's choice of insurer either because the insurance available from that nominated insurer is unsatisfactory or because the premiums payable are excessive.

Managing agents

Section 30B entitles a recognised tenants' association to serve notice on the landlord requesting him to consult the association on matters relating to the appointment of a managing agent. Where such notice is served before the landlord has appointed a managing agent, the landlord must, before making such appointment, serve notice on the association specifying:

- the name of the proposed managing agent
- the obligations which the agent will undertake for the landlord
- a period of at least one month within which the association may make observations on the proposed appointment.

Where a managing agent already exists when the association's notice is served, the landlord must within one month serve a response specifying the agent's obligations and a reasonable period within which the association may make observation on the manner in which the managing agent has been discharging those obligations and on the desirability of his continuing to discharge them.

Where a managing agent has been appointed and the association have served notice, the landlord must at least once every five years serve notice specifying any changes which have occurred since the last notice and a reasonable period during which the association may make representations. A fresh notice must also be served if the landlord proposes to appoint any new managing agent.

Statements of account

Section 21 of the Landlord and Tenant Act 1985 (as amended) requires ground landlords to supply to each leaseholder an annual statement of account relating to service charges, the aggregate amount standing to the credit of leaseholders and related matters. That statement of account must be supplied to each leaseholder within six months from the end of the relevant accounting period. That statement must be accompanied by a certificate of a qualified accountant to the effect that the account deals fairly with the matters to which it is required to deal and is supported by accounts, receipts and other relevant documentation and also summarising the rights and obligations of leaseholders in relation to service charges.

Section 21A allows a leaseholder to withhold payment of a service charge if the relevant statement of account has not been provided within the six month period or if the form or content of that document does not comply with statutory requirements. The maximum amount which the leaseholder may withhold in those circumstances is an amount equal to the aggregate of service charges already paid by him in relation to that accounting period and so much as currently stands to his credit. Payment of the backdated amount will immediately full due if either:

- the relevant statement of account is subsequently supplied in a form which meets statutory requirements or
- if an LVT determines that the landlord has a reasonable excuse for the failure to comply with the statutory requirements.

A leaseholder will never be obliged to pay interest to a landlord in circumstances where he has been entitled to refuse or delay payment of those service charges. Section 21B requires any demand for service charges to be accompanied by a summary of the rights and obligations of the leaseholders in relation to those charges. The form and content of that summarised information may be set out in regulations. Again a leaseholder may withhold payment of a service charge if these provisions are not complied with.

Section 22 of the 1985 Act (as amended) allows any leaseholder to insist on reasonable facilities to inspect accounts, receipts or any other relevant documents relating to the service charges or to take copies of or extracts from such documents. But leaseholders cannot exercise this right more than six months after the date before which the landlord is required to provide the statement of account or (if later) six months from the date the statement was actually supplied.

Other service charge provisions

Section 20B states that a demand for service charges is only binding on a leaseholder if it is served on the leaseholder within 18 months of the date those costs were originally incurred unless within that 18 month period the leaseholder has been notified in writing that those costs have been incurred and that he would subsequently be required to contribute towards them.

Section 25 makes it an offence (punishable by a fine of up to £2,500) for a landlord to fail, without reasonable excuse, to perform any duty imposed on him.

Section 26 makes clear that a public authority is only required to comply with the statutory requirements relating to service charges in relation to residential leases granted for more than 21 years.

Challenging service charges

Section 19 states generally that service charges are only recoverable from residential leaseholders to the extent that they are reasonably incurred and only if the services or works to which they relate are of a reasonable standard.

Section 27A allows either a landlord or a leaseholder to apply to an LVT for a determination whether a service charge is payable and, if so, as to the:

• person by whom it is payable

- person to whom it is payable
- amount payable
- date on which it is due and
- manner in which it is to be paid.

Section 27A(2) preserves a leaseholder's right to challenge a service charge even if payment has already been made — so long as that liability has not already been agreed or admitted or determined by a court or arbitrator (pursuant to a post-dispute arbitration agreement). This enables payment to be made "under protest" without prejudice to the leaseholder's right to question the principle or amount of the demand. But a service charge cannot be challenged if it has previously been agreed or admitted by the tenant or adjudicated as being due by a court (or by an arbitrator pursuant to a post-dispute arbitration agreement — see Chapter 14).

Section 27(3) also allows a landlord to apply for a determination in respect of proposed works or services which may become the subject of a future service charge if they were to be commissioned.

Relevant cases on service charges

Martin v *Maryland Estates Ltd* [1999] 2 EGLR 53 — where the Court of Appeal disallowed the costs to a ground landlord in carrying out additional works which had not been the subject of a statutory consultation exercise. The need for those additional works was discovered during the carrying out of the original works, which had been the subject of full consultation.

However the cost of those additional works was almost as much as the cost of the original scheduled works and the landlord made a deliberate decision not to consult the leaseholders in respect of them. The landlord was already aware that the leaseholders were unhappy with the original scheduled works, even though they had been consulted about them. Blofeld J said that a reasonable landlord would in those circumstances have informed leaseholders of the extra work it was proposed to carry out. The appeal judges upheld the decision of the trial-judge to refuse to dispense with the statutory consultation requirements in relation to the additional works.

Parker and Beckett v *Parham* (LTL 11/7/2003) — the Lands Tribunal ruled that it was not reasonable for ground landlords to recover as advanced service charge the cost of works which they had not yet carried out where it was uncertain when those works would be carried out. The Lands Tribunal also prevented the landlords from recovering from the leaseholders the costs in initiating proceedings in the LVT to certify the reasonableness of proposed service charges, where such proceedings were regarded only as a management tool to ensure payment of service charges. The LVT had previously formed the view from the parties conduct in the proceedings that it was appropriate to disallow the landlord's costs. However the leaseholders' application to spread the cost of service charges over three years failed as such an order would be beyond an LVT's jurisdiction.

Administration charges

Administration charges are the costs which a ground landlord is contractually entitled to charge in connection with any of the following landlord duties:

- in dealing with any request by the leaseholder for a specific approval under his lease (such as for proposed alterations) — including the cost of documenting such approval (if granted)
- in providing information or documents requested by the leaseholder
- in following up overdue payments under the lease
- arising out of any breach (or alleged breach) of a covenant or condition in the lease.

As a matter of general law, ground landlords can only charge what is reasonable for dealing with these matters. Such costs are mainly intended to cover reimbursement of the landlord's legal and other professional costs incurred in complying with a tenant's request or taking necessary action. But until now there has been no convenient mechanism by which leaseholders can challenge the reasonableness of such administration costs. Those matters are now dealt with in schedule 11 to the CLRA.

Clause 1 (3) of that schedule defines administration charges as "variable" if they are not fixed by the terms of the lease or calculated in accordance with some other formula specified in the lease. An example of a "fixed administration charge" is the nominal sum which an incoming leaseholder is normally required to pay a ground landlord to cover his time and trouble in recording details of the new leaseholder.

Clause 2 states that a "variable" administration charge can be recovered only to the extent that it is reasonable.

Clause 3 enables a leaseholder to apply to an LVT to vary the terms of a lease on the grounds that the amount of a specified administration charge is unreasonable or any formula specified to calculate that amount is unreasonable.

Clause 4 requires any demand for the payment of an administration charge to be accompanied by a summary of the rights and obligations of leaseholders in relation to those charges. Regulations may state the required form and content of those summaries. Clause 4 (3) enables a leaseholder to withhold payment of an administration charge until a demand has been made which complies with the relevant statutory formalities.

Clause 5 enables a leaseholder to apply to an LVT for a determination whether an administration charge is payable and, if it is, as to:

- the person by whom it is payable
- the person to whom it is payable
- the amount which is payable
- the date at or by which it is payable and
- the manner in which it is payable.

As with service charges (see above) a challenge can be made to an LVT even if payment of the administration charge has already been made. It is because ground landlords will commonly require payment of administration charges up front before they will deal with a leaseholder's request or provide information.

Service charge checklist

- Is it something which the terms of the lease allow the landlord to recover?
- Has the landlord complied with the Service Charges (Consultation Requirements) (England) Regulations (where applicable)?
- Have annual statements of account been provided (section 21)?
- Was the leaseholder notified of the service charge item within 18 months from the date the costs were incurred (section 20B)?
- Has the amount of the service/administration charge been admitted or adjudicated upon?
- Are there grounds to challenge the reasonableness of the charge?

The Right to Manage

There are several ways in which leaseholders can collectively acquire the right to manage the block in which their individual flats form part.

Acquiring the right to manage is not the same as buying out the landlord's interest. Leaseholders will continue to pay ground rent and will continue to be answerable to the ground landlord for compliance with the conditions of the lease.

Acquiring the right to manage means that leaseholders can collectively take control of their own affairs. It will then be leaseholders that will collectively arrange buildings insurance for the block and collectively make decisions concerning the repair and maintenance of the block, who is to be contracted to carry out those repairs and maintenance, and how much that work will cost. However taking control of management will not absolve the leaseholders' nominee from having to comply with the detailed requirements set out in Chapter 5 requiring consultation over potential service-charge items and to comply with all other regulatory requirements relating to the levying of service charges. As well as taking over control of their own affairs, the leaseholders will also take on additional responsibility. The new management organisation will have a duty not only to every leaseholder but also to the ground landlord, whose interest will retain some monetary value.

Management companies

In granting new leases, some residential ground landlords already delegate maintenance to a separate management company specifically set up for that purpose. Each leaseholder will then be required to take membership of that company and to transfer their membership when the lease is sold. The management company will be party to each residential lease and will contract with each leaseholder to manage the block and levy service charges. Membership of the organisation will ensure that each leaseholder has an opportunity to participate in each management decision which is made. Communal areas, such as car parks, corridors, lifts, as well as the structural parts of the building, may also be leased directly to the management organisation. By setting up matters in this way, the landlord's interest becomes a pure financial investment, absolved from any responsibility for repairs, maintenance, insurance or management of the building. The arrangement may also add value to the individual flats comprised within the building by guaranteeing leaseholders control of the day to day running of their block.

The idea that the courts could unilaterally taking away a ground landlord's rights and responsibilities regarding repair, insurance and management of a building and transfer those responsibilities to a leaseholder's nominee first arose in the case of *Hart* v *Emelkirk Ltd* [1983]2 EGLR 41, when the leaseholders of 11–20 Cambridge Mansions, Wandsworth, persuaded Goulding J to approve the appointment of a chartered surveyor to collect rents and service charges and exercise the landlord's powers of management, in circumstances where the landlord had failed to collect rent and maintain the building and prevent deterioration to the premises.

Four years later the courts' common law power to appoint a receiver to manage a ground landlord's affairs became enshrined in statute and is now contained in Part II of the Landlord and Tenant Act 1987, which still remains in force as amended.

Where landlord in default

Sections 21–24 of the Landlord and Tenant Act 1987 (as amended by the Housing Act 1996 and the CLRA 2002) enable *any* residential leaseholder to apply to an LVT for an order appointing a manager to act in relation to that building. There are of course some statutory exceptions. It will not apply where the landlord is himself resident in the building unless at least one half of all flats contained within the building are held on leases of more than 21 years. Neither would it apply to certain "exempt" landlords or certain charity lands. Subject to this a single leaseholder or more than one acting together may make an application to the LVT.

Before applying for any order under these provisions, the applicant leaseholder must first serve a preliminary notice on the landlord and any separate management organisation which is party to the lease, which must (via section 22):

- specify the applicant's name and address
- state that the applicant intends to apply for an order under section 24 to an LVT but that he will not do so if the stated matters can and are remedied by the recipient under notice within such reasonable period as may be specified
- specify the grounds on which the LVT will be asked to make the order and the matters on which the applicant will rely to establish those grounds.

Where the landlord's interest is itself mortgaged, it is the landlord's responsibility to copy the notice to the mortgagee.

Following service of the preliminary notice, section 23 prevents any application to an LVT being made until the period specified in the preliminary notice has expired without the recipient having taken appropriate steps to have complied with it.

If the response to the preliminary notice is negative (or if there is no response) and if the problem is not otherwise resolved, section 24 of the Act then allows the applicant to apply to an LVT for an order appointing a manager to carry out specified management functions in place of the landlord.

The LVT can only make such an order when it is "just and convenient" to make the order in all the circumstances of the case; and provided that at least one of the following grounds have been proved by the applicant:

- that the landlord or management company is in breach of any management obligation owed to the leaseholder under his tenancy
- that unreasonable service charges have been made or are proposed or are likely to be made
- that unreasonable variable administration charges have been made or are proposed or likely
- that the landlord or management company has failed to comply with any relevant provision of a code of practice approved by the Secretary of State under Section 87 Leasehold Reform Housing and Urban Development Act 1993 (Codes of Management Practice) or

- where other circumstances exist which make it just and convenient for the order to be made.

Management is defined by section 24 (11) to include references to the repair, maintenance, improvement or insurance of the block. Once an order has been made under section 24, it is also important that the appropriate notification of that order is given to HM Land Registry (if the landlord's interest is registered) or under the Land Charges Act 1972 (if a landlord's interest is unregistered). This is to alert any third party intending to deal with the landlord's interest of the existence of the order.

At the time of writing, there is only one significant court decision relating to the above provisions. That is *Taylor* v *Blaquiere* [2003] 1 EGLR 52, when the Court of Appeal defined the role and responsibilities of a manager appointed by an LVT.

Hugh Blaquiere was a leaseholder of Flat 2, Nos 14, 15 & 16 Hyde Park Gardens, London W2. His flat comprised a large semi-basement extending under the terraces of other flats. Mr Blaquiere leased his flat from Hyde Park Estates (Guernsey) Ltd, which in turn leased from the Church Commissioners.

Under the sublease with Mr Blaquiere, Guernsey were obliged to insure the premises and maintain it in good structural repair and condition, decorating the exterior as and when necessary. Unfortunately, however, that company was in an administrative receivership and was no longer undertaking those obligations. As a result, Mr Blaquiere's flat suffered from a range of problems including ingress of water, timber-rot, drainage difficulties, general disrepair and lack of decoration. According to Mr Blaquiere, those defects made the flat uninhabitable. Mr Blaquiere and another leaseholder jointly applied to an LVT to appoint a manager. That application succeeded and on 1 December 1998, Mr Maunder Taylor of 1320 High Road, Whetstone, London N29 was appointed manager under the 1987 Act. But Mr Blaquiere's problems did not end there.

Shortly after his appointment, Mr Maunder Taylor submitted a draft service charge budget for the forthcoming year, which included major building works. He discussed these proposals with the leaseholders, tendered those works and engaged contractors on 12 January 1999. The works were completed during the course of that year. Mr Maunder Taylor initially assessed Mr Blaquiere's share of those costs at £24,000. He then claimed reimbursement for further expenditure, which increased Mr Blaquiere's share to more than £62,000.

Mr Blaquiere then sought to set-off against those costs a claim for damages, which he said were due to him from landlords Guernsey consequent on their previous failures to carry out their repair and maintenance obligations. The question for the Court of Appeal was whether the law allowed such a set-off, even if the leaseholder had a valid claim against the landlord. The court ruled that there could be no set-off against a court appointed manager.

It was ruled that the purpose of Part II Landlord and Tenant Act 1987 was to provide a scheme for the appointment of a manager who would carry out the functions required by the LVT. An appointed manager did not act as the landlord's agent. He acted in his own right as an LVT appointed official. Mr Maunder Taylor's claim was therefore brought in his capacity as manager and entirely on his own account.

It is also worth noting the comments of the trial judge Recorder Hamlin, against whose decision the appeal was lodged:

> The purpose of Part II is to enable tenants to outflank an irresponsible landlord and to retain a skilled professional to enable the building to be put back into repair. That professional could not be regarded as a guarantor of the landlord.

The new right to manage

The drawback with seeking a transfer of management under Part II of the 1987 Act is the need to prove that the ground landlord has completely disregarded its maintenance and management responsibilities. Small deficiencies in service provision may not be sufficient to convince an LVT that it is "just and reasonable" to transfer management. And even if management is transferred to a professional manager, the case of *Taylor* v *Blaquiere* shows that individual leaseholders may still be dissatisfied with the outcome.

Part II of the CLRA introduces an alternative means by which a majority of leaseholders can collectively take over the management of a block without having to prove fault on the part of the existing landlord or having to approach any court or tribunal (unless the exercise of the right is disputed). This alternative "right" is entirely process-driven, which means that so long as the eligibility criteria are satisfied and the correct procedures followed, management will transfer automatically. The value of the landlord's interest will remain unaffected, but leaseholders will thereafter be able to make their own management decisions on a democratic basis. The procedural checklist can be summarised as follows:

- preliminary information gathering to assess whether eligibility criteria can be met
- set up a right to manage (RTM) company in the statutory format
- issue notice of participation to other leaseholders
- issue claim notice
- receive landlord's counternotice
- disputes resolution (if applicable)
- third party contractual issues
- notify Land Registry
- commence management.

There are also other issues which leaseholders would be wise to consider before initiating the statutory procedure:

- Why take over management? If the landlord is already managing the block responsibly, professionally and cost effectively, there maybe no advantage to the leaseholders in taking over management. There would certainly be additional responsibility and possibly additional expense involved for the leaseholders.
- Do the leaseholders have the expertise — and more importantly the time — to deal with their own management? If the answer is "no" or "doubtful", those management functions would, in turn, have to be contracted out to professional managing agents. However, even with such contracting out, it will still be necessary for someone to take charge of the RTM company and ensure that its meetings and decisions are correctly organised and minuted and that statutory returns and accounts are lodged with Companies House within the statutory time-limits and in the correct manner.
- How much will it cost to take over management? Professional expertise will be required to assess eligibility, set up the correct company and draft and issue the correct notices. Professional help may also be needed to deal with the associated contractual issues, which are likely to arise in relation to larger blocks of flats. Leaseholders will also be required to reimburse the landlord's necessary and reasonable costs in responding to the leaseholders' management claim. If the leaseholders' claim is successfully challenged by the landlord, the leaseholders will also be required to pay the landlord's costs arising out of the LVT hearing.

Eligibility requirements

Section 72 of the Commonhold and Leasehold Reform Act 2002 states that the right to manage will apply to a building containing two or more flats held by qualifying tenants, where the total number of flats held by such tenants is not less than two thirds of the total number of flats contained in the building. However the following types of building are excluded by schedule 6 from this right:

- where more than 25% of the internal floor area of the building (excluding the common parts) is used for a non-residential purpose (eg commercially)
- where a building (other than a purpose-built block of flats) comprises not more than four residential units and at least one of those units is the landlords own home (or the home of their spouse, parent, child or in-law)
- local authority housing
- where a block is already being managed by a RTM Company or has ceased to be so managed within the previous four years (unless an LVT orders otherwise).

Section 75 defines a qualifying tenant as being a tenant of a flat under a long lease — which is defined in turn by section 76 to mean a lease for a term exceeding 21 years. The definition also includes shared ownership leases.

Establishing an RTM company

Having made the initial appraisal, the first formal step is to set up an RTM company, which complies exactly with the requirements of the legislation. Section 73 (2) states that a company is an RTM company only if:

- it is a private company limited by guarantee and
- its memorandum of association states its object, or one of its objects, is the acquisition and exercise of the right to manage the premises.

Section 74 goes on to say that the only persons who are entitled to membership of an RTM company are qualifying tenants of flats contained in the premises and relevant landlords. The memorandum and articles of association jointly comprise the constitution of the RTM company and these must conform to the RTM Companies (Memorandum and Articles of Association) (England) Regulations 2003.

Limited by guarantee means that the company will not have a share capital and that those participating in it will be members not shareholders. However it is still a limited liability organisation and the liability of each member will be limited to £1. The fact that a company is limited by guarantee generally signifies that it is a "not for profit" organisation set up for a purpose other than for trading.

Part 1 of the schedule to the regulations set out the memorandum for every RTM company. It is this memorandum which sets out the powers of the company.

Article 1 of the memorandum requires the name of company to include the words RTM Company Limited. Article 4 (g) sets out the principal function of the company which is:

> To provide and maintain services and amenities of every description in relation to the premises; to maintain, repair, renew, redecorate, repaint and clean the premises; and to cultivate, maintain, landscape and plant any land, gardens and grounds comprised in the premises.

To achieve this, article 4(h) allows the company:

> To enter into contracts with builders, decorators, cleaners, tenants, contractors, gardeners, or any other person, to consult and retain any professional advisors and to employ any staff and managing or other agents; and to pay, reward or remunerate in any way any person supplying goods or services to the company.

Article 4 (l) requires the company:

> To insure the premises or any other property of the company or in which it has an interest against damage or destruction and such other risks as maybe considered necessary, appropriate or desirable and to insure the company and its directors, officers or auditors against public liability and other risks which it may consider prudent or desirable to ensure against.

Other provisions in the memorandum empower the company to make the initial application for management; exercise landlord's management functions; monitor and enforce tenants' lease covenants; take or defend legal proceedings; collect service charges; maintain a sinking fund; lend and advance money on credit; borrow and raise money; award pensions; and do everything else which is relevant to the company's objective. Article 8 states that if on dissolution of the company there is any surplus balance, that is to be shared with the existing membership of the company at that time.

The articles of association deal with procedural matters relating to membership, the voting and the appointment and responsibilities of directors. Articles 4 and 5 restate the legal position that only qualifying tenants and landlords can participate in the company. Although qualifying leaseholders are entitled to membership, there is no compulsion and article 12 entitles any member to withdraw from the company on giving a week's notice. However, members may not withdraw during the period from issue of a claim notice (see below) until the date the right to manage takes effect or is withdrawn.

Any qualifying leaseholder who does not take up membership of the company or who subsequently withdraws, will be unable to participate directly in management decisions but will retain their statutory rights to be consulted on proposed service charge items and to challenge their reasonableness.

Articles 14 to 48 deal with the day to day business of the company and how decisions are to be made. Articles 14 to 21 deal with the calling of general meetings — and articles 22 to 36 deal with the proceedings at those meetings. Article 23 states that no business can be transacted at a meeting unless a quorum of at least 20% of the membership is present or represented. Where a quorum has not assembled within half an hour from the scheduled start of a meeting, that meeting is to stand adjourned for one week. Decisions of the company will be made on a show of hands unless a ballot is demanded by the chairman, at least two voting members, or a member holding not less than one tenth of total voting rights.

The issue of voting rights is complex. If the membership comprises only qualifying leaseholders, then there is one vote for each qualifying flat (article 38).

If landlords are represented within the company, article 29 states that each qualifying flat will have one vote for each landlord. There may be more than one landlord if, for instance, the immediate landlord is himself a leaseholder. This is explained in the following example.

Example — calculation of voting rights

Station Court is divided into six flats, each held on the balance of a 99 year underlease at a fixed annual ground rent of £5. The qualifying leaseholders are: T1, T2, T3, T4, T5 and T6. The immediate ground landlord of those underleases is L1. However, L1 is himself a leaseholder of the whole building from superior landlord L2 under a 999 year lease at a fixed annual rent of £1.

The qualifying leaseholders have set up Station Court RTM Company Ltd to acquire management of Station Court. Under article 5, all the qualifying leaseholders and both L1 and L2 are entitled to membership of the company. Therefore applying the formula set out in article 39, each of the qualifying leaseholders will be entitled to two votes. Under article 39(f), L1 and L2 are each entitled to one vote. This formula guarantees that qualifying leaseholders will always be in the majority when important management decisions are taken — provided those leaseholders take the trouble to attend meetings and vote. But supposing T6 later sells his flat to L2?

As well as remaining the superior landlord of the whole of Station Court, L2 will also be the qualifying leaseholder of Flat 6. Therefore L2 would be entitled to membership of the company both as a qualifying tenant and as a superior landlord. Using the same formula, one would therefore assume that L2 will have three votes in aggregate (two as qualifying leaseholder and one as landlord).

To make the calculation a little more complicated, assume now that on the ground floor of Station Court is a small lock-up shop of equivalent size to one of the qualifying flats. The size of the commercial unit is too small to prevent the leaseholders exercising their right to manage, but its existence will have to be reflected in the voting pattern. Under article 39(b) a proportionate number of votes must be allocated to the non-residential part, having regard to a comparison of its total internal floor area with the aggregate of the residential parts. As the commercial unit is an equivalent size to a single qualifying flat, under that formula the commercial unit would attract two votes, which under article 39(d) would belong to the immediate landlord, L1. Thus the total voting pattern will be: Ts1–5 (two votes each); L1 (two votes for the shop and one vote as landlord); L2 (two votes as qualifying tenant of flat 6 and one vote as superior landlord). Thus there will be 16 votes in aggregate. Any commercial tenant of the lock-up shop cannot participate in an RTM company or have any say in the decision making process. The complexity of the voting system may itself provide a fruitful source of litigation. Article 42 requires that any objection or query as to anyone's voting entitlement must be raised in the first instance with the chairman of the particular meeting, whose decision will be final so far as that meeting is concerned.

Articles 51–82 deal with the appointment, powers and responsibilities of directors. Although there must be a members' AGM for every company, most of the day to day running of the company is dealt with by its directors, who can make decisions on behalf of the wider membership. For a small block of flats like Station Court it is probable that each of the members of the company will also be its directors. For a larger block containing many units, it would be impractical to convene a meeting of the entire membership each time a series of decisions need to be made. It is therefore more likely that the members of such a company would want to nominate a few directors from among their membership to take decisions on their behalf.

Notice of participation

Following the formation of an RTM company, the second formal step to be taken by the initiators of the procedure is to serve a notice of participation. This is a notice to be served by the initiators on every qualifying leaseholder who is not already a member of company, giving them an opportunity to take up membership of the company.

Section 78 of the CLRA states that before claiming the right to manage, an RTM company must first give formal notice to every other qualifying tenant who is not already a member (or prospective member) of the RTM company. Note that the notice must come from the RTM company — not anyone else. The contents of the notice inviting participation are set out in section 78(2) and article 3 of the Right to Manage (Prescribed Particulars and Forms) (England) Regulations 2003. The specimen form of notice of invitation to participate is also set out in Schedule 1 to those regulations. Among the information to be included in that notice are:

- the names of the existing members of the company, its directors and the company secretary
- the names of the landlord(s) and of anyone else who is party to a lease otherwise as landlord or tenant
- what the company will be responsible for and what limitations apply as regards its powers of management
- whether the company intends to employ a managing agent, and if so, the name and address of that managing agent
- the fact that members of the company will be responsible for the landlord's costs once a claim notice is issued
- inviting the recipient of the notice to become a member of the company.

The participation notice must also be accompanied by copy of the memorandum and articles of association of the RTM company or state when and where those documents can be inspected and the cost of the obtaining a copy. Note that at this stage of the process the landlord is not involved.

Service of claim notice

The service of the claim notice under section 79 CLRA is the stage at which the RTM process becomes potentially contentious. But before a claim notice can be served the following criteria must first be satisfied:

- a minimum of two thirds of the total number of flats in the building must be held by qualifying tenants and none of the statutory exemptions must apply
- a participation notice must have been served on every qualifying tenant (who is not already a member or prospective member of the company) at least 14 days in advance
- at least half the qualifying tenants must have taken up membership with the RTM company.

Provided each of these criteria are satisfied, the RTM company can then proceed to serve its claim notice on the ground landlord and anyone else who is party to the lease (such as an existing management company). The copy of the notice must also be given to every qualifying tenant as well as any manager appointed under Part 2 of the 1987 Act (see above). However, the fact that someone cannot be found does not negate the process so long as the notice can be served on other people.

Section 80 and the Right to Manage (Prescribed Particulars and Forms) (England) Regulations 2003 set out the contents of a claim notice and the form it must take. A pro-forma claim notice is set out in schedule 2 of the regulations.

The notice, which must be given by the RTM company, must give basic information about itself and give the full names of everyone who is both the qualifying tenant of a flat contained in the premises and a member of the company, including the address of his flat. In relation to each named person, the notice must also state the date on which their lease was entered into, the term for which it was granted, the date the term commenced and such other information as is necessary to identify the lease. The notice must also state the landlord's right to give a counternotice within a period of not less than one month. The notice must also state the date the company intends to acquire the right to manage, which must be at least three months after the deadline for the landlords counternotice. This is the date the formal handover the management will take place if the notice is not challenged by the landlord. A landlord who does not intend to dispute the company's entitlement to manage is also called upon to notify the company of any management contracts relating to the block (see below). The notice must also inform landlords of their own rights to take up membership of the company. Most crucially, the claim notice must accurately set out the grounds on which it is claiming the right to manage.

Any claim notice must be carefully drafted to ensure that it accurately reflects the situation and professional advice sought where appropriate. However, it is clear from section 81(1) of the CLRA that minor errors will not invalid the notice so long as its meaning is clear and no one is prejudiced by that error.

Note also section 82, which allows an RTM company to demand information in anyone's possession or control which the company reasonably requires to ascertain the details required to complete a claim notice. Where that information is contained within a document in that person's possession or control, the RTM company may demand to inspect that document. Any such demands must be complied with within 28 days.

Note also that once a claim notice has been given, anyone who is a landlord or party to a lease (otherwise than landlord or tenant) has the right to demand access to any part of the building (including individual flats) in connection with any matter arising out of the claim to acquire the right to manage. The occupier of that part of the building is entitled to at least 10 days notice before access can take place. It is clear that many landlords will want to inspect the whole of the premises the subject of a claim notice to satisfy themselves that, in physical terms, the criteria for establishing a right to manage had been met.

Landlord's counternotice

Under section 84 of the CLRA anyone receiving a claim notice has at least one month in which to serve a counternotice on the RTM company. The deadline for serving the counter-notice will be contained within the claim notice itself. For this purpose a counter-notice is a notice containing a statement either:

- admitting that the RTM company was on the relevant date, entitled to acquire the right to manage the specified premises or
- alleging that, by reason of a specified statutory provision, the RTM company was not on that date so entitled.

The counternotice must also contain information required by article 5 of the Right to Manage (Prescribed Particulars and Forms) (England) Regulations 2003 and conform to the proforma set out in schedule 3 of those regulations. If the landlord serves a counternotice admitting the right to manage (or if the landlord fails to serve any counternotice at all), management will transfer to the RTM company on the date specified in the claim notice. If the landlord does dispute the right to manage, his grounds for doing so must be set out in the counternotice. If a counternotice is served disputing the right to manage, management will not transfer unless the RTM company successfully applies to an LVT for a determination that it was on the relevant date entitled to acquire the right to manage the premises. That application to the LVT must be made within two months from receipt of the counternotice. Possible grounds on which a right to manage might be disputed are:

- less than two thirds of the flats are held by qualifying tenants
- one of the statutory exemptions applies — eg if more than 25% of the internal floor area is occupied commercially
- the constitution or membership of the RTM company do not satisfy statutory requirements for an RTM company
- the notice inviting participation has not been served as required by law
- less than 50% of the qualifying tenants were members (or prospective members) of the RTM company on the date the claim notice was served.

Untraceable landlord

Where an RTM company has been established and has issued a participation notice and has satisfied other statutory criteria but cannot serve a claim notice because the whereabouts of any of the intended recipients are untraceable, section 85 of the CLRA allows the RTM company to apply to an LVT for an order that the company is acquire the right to manage. Notice of such application must have been given to every qualifying tenant in the premises. An LVT may also require the company to take further steps by way of advertisement or otherwise in an attempt to trace landlords or anyone else who is party to such leases otherwise than as landlord or tenant. If that person is traced, proceedings before the LVT will be halted and the process will continue as if the RTM company had served a valid claim notice on that person.

Costs

Mention has already been made of the fact that an RTM company who serves a claim notice will be responsible for the landlord's costs arising out of it.

Section 88 of the CLRA makes an RTM company liable for the reasonable costs incurred by a landlord or other party to a lease (other than the tenant) in consequence of a claim notice given by the company relating to the premises whether that notice succeeds or fails. Any costs incurred by that person in respect of professional services are regarded as reasonable only to the extent that the costs of such services might reasonably be expected to have been incurred by him if he had been expected to pay those costs himself. An RTM company will only be responsible for a landlord's LVT costs if the landlord's objection is upheld. The amount of any costs payable by the RTM to another party will be assessed (in default of agreement) by the LVT.

Section 89 also makes an RTM liable for any landlord's costs if a claim notice is either withdrawn or if it lapses because it is disputed and the RTM fails to apply to an LVT for a determination within the statutory two month period. Section 89(3) also makes clear that everyone who is or who has been a member of the RTM company can be held personally liable for the whole of those costs. This is one of the risks which any participant in an RTM company will take. Note section 86 which allows an RTM company to withdraw a claim notice at anytime before it acquires the right to manage by issue of a notice of the withdrawal. The RTM company will then be liable for reasonable costs incurred by the landlord up to the date of that withdrawal. If at any stage of the proceedings it becomes apparent that the RTM company's claim has no realistic prospect of success, notice of withdrawal should be issued as soon as possible before unnecessary additional costs are incurred.

Treatment of existing contracts

Unfortunately the Act skates over the contractual complications which can arise when management of a large block of flats is transferred from a ground landlord to an RTM company. All that the law requires the ground landlord to do is alert the RTM company and existing contractors to the existence of such contracts and the pending transfer of management and leave it up to the RTM company and contractors to make their own arrangements for the continuation of the contracted service or otherwise. There is no automatic transfer of such contracts from the ground landlord to the RTM company. Neither does statute provide any convenient disputes resolution procedure in the event that such contractual issues cannot be easily resolved.

Management contracts are defined by section 91(2) of the CLRA to comprise any contract between the existing manager of the property (which may be the ground landlord) and a contractor, by which a contractor has agreed to provide services or do any other thing relating to something which will be a function of the RTM company once management is transferred. Therefore a management contract could include such things as grounds maintenance, janitorial services, window cleaning and administrative functions.

Section 92 makes it the responsibility of the existing manager party to give a Contractor Notice to the existing contractor and a "contract notice" to the RTM company. Those notices must be given on or as soon as possible after the determination date. This is the date on which the RTM company's statutory right to take over management has become established in law, either because the landlord has not served a counternotice (see above) or when the RTM company succeeds in its claim before an LVT. Where that contract is entered into after the determination date but before actual handover of management, such notices must be given on or as soon as possible after the date the contract is entered into. A contractor notice must:

- identify the relevant contract
- state that the right to manage the premises is to be acquired by an RTM company and state the name and registered office of that company
- specify the handover date and
- that should the contractor wish to provide similar services to the RTM company it should contact the RTM company at the address stated [Regulation 6 of the Right to Manage Prescribed Particulars and Forms (England) Regulations (2003)].

Any contractor receiving a contractor notice must copy it to any of its subcontractors and give the RTM company notice in respect of that subcontract. A contract notice must give particulars about the contract to which it relates and of the relevant contract or subcontractor (including the contractor's address) and also state that should the RTM company wish avail itself of the contractor's or subcontractor's services, it should contact the relevant contractor or subcontractor at the address stated (a requirement of article 7 of the regulations).

Associated handover matters

Once the right to manage has been legally established, section 93 of the CLRA enables the RTM company to give notice to the ground landlord or other existing manager, requiring it to provide the RTM company with any information in its possession or control which the company reasonably requires in connection with the exercise with the right to manage. This includes the right for an authorised representative of the RTM company to inspect any document containing relevant information and to be provided with a copy of that document. However the landlord or existing manager cannot be required to act on such requests until after the actual handover of management. However, once management has been handed over, the landlord or existing manager will only have 28 days from the date of the original issue of the notice to comply.

Section 94 requires the ground landlord or other existing manager to hand over to the RTM company a sum equivalent to any uncommitted service charges on the date of handover (known as the acquisition date). Those uncommitted service charges must equate to the aggregate of sums actually paid by leaseholders together with the proceeds of any investment of those sums — but deducting such amount as is required to meet costs incurred before the acquisition date in relation to matters for which the service charges were payable. Where there is a dispute over the amount to be handed over in relation to uncommitted service charges, either the landlord (an existing manager) or the RTM company may apply to an LVT to resolve the issue.

To illustrate the complexities of the type of contractual issues which may need to be resolved when management of a large number of flats is transferred, consider the following fictitious example:

Example

Melony Mews is a block of 100 residential flats, 99 of which are let on long residential leases. The remaining flat is occupied by a 61-year-old caretaker who provides security and janitorial services.

The freehold is owned by Squirrel Investments Ltd. Squirrel has contracted out its management functions to Northside Management Ltd on a three-year contract, of which there are two years left to run. Squirrel employs the caretaker directly and provides the caretaker's flat.

Stevens Maintenance Partnership cut the grass on a fortnightly basis and clean the windows monthly. Their contract runs indefinitely until terminated by either side on three month's notice. Squirrel has complied with the requirements of the Landlord and Tenant Act 1985 (as amended) in relation to the letting of these contracts (see Chapter 5).

Six of the most active leaseholders have got together to form Melony RTM Ltd to take over management of the block. In response to a participation notice served on the remaining 93 leaseholders, 60 of them have taken up membership of the RTM company. A notice claiming the right to manage was subsequently served by the RTM company on Squirrel. Squirrel did not serve any counternotice disputing the right to manage and the formal handover of management will take place on a specified date four months hence. Squirrel has served contractor notices on Northside and Stevens Partnership notifying those organisations of the pending transfer of management and providing the name and address of Melony RTM Ltd. It has also served contract notices on the RTM company giving notice of its contractual arrangements with Northside and Stevens Partnership together with the names and addresses of those organisations. However the position as regards the caretaker is more complex as he is not a contractor of Squirrel but a direct employee. We now look at the position regarding each of these contractual arrangements.

- *The contract with Northside*
  As stated, this is a fixed term contract with two years left to run. That contract contains no provisions for termination in the event of any transfer of management. However if Northside is willing to provide services for Melony RTM Ltd and if Melony is willing to purchase those services, there could be a novation of that contract whereby the RTM company would (by agreement) step into the shoes of Squirrel and continue purchasing the same services on its own behalf, with or without variation. But if either Melony or Northside do not wish to continue purchasing (or as the case may be supplying) those services, a complex contractual situation would arise. This is because (technically at least) Squirrel and Northside would (unless otherwise agreed) remain contractually committed to each other for the residual two years of the contract, even though there is nothing for Northside to manage.

- *The situation regarding Stevens Partnership*
  As this contract can be terminated by either side on three months notice no contractual problem will arise if Stevens and the RTM company cannot agree terms which will enable Stevens to continue providing the service.

- *The situation regarding the caretaker*
  Both Squirrel and the RTM company would be wise to seek the advice of an employment law specialist as under the Transfer of Undertakings (Protection of Employment) Regulations 1981 it is likely that the caretaker's contract of employment might transfer automatically from Squirrel to the RTM company on the same terms and conditions as he previously enjoyed. It is also possible that any claims or potential claims which the caretaker had

against Squirrel would carry over to his new employer, the RTM Company. His situation is complicated further by the fact that he occupies one of the flats rent free as part of his employment package. The RTM company would have to ensure, in its handover arrangements with Squirrel, that it would continue to have use of the flat for its resident caretaker. The terms and conditions of that use are themselves a potential source of dispute between a ground landlord and an RTM company until the courts have stated their position on such issues. The fact that the ground landlord retains ownership of one of the flats will also give Squirrel an additional vote at the RTM company, assuming it exercises its right to membership of that organisation.

Exercising the right to manage

Chapter 15 deals with the practical issues which will inevitably arise in the running of a management company. However, even after management has been acquired, the RTM company will *not* take over *all* the landlord's rights and responsibilities in relation to the management of the block. The RTM company will still have to consult the ground landlord on strategic matters relating to alterations (or in some cases assignment or subletting), for which the landlord will retain a limited veto. The RTM company will also have to account to the ground landlord in the way it carries out it functions and monitor compliance with leasehold covenants. Although the RTM company will have the right to take legal proceedings to recover service charges and enforce compliance with other lease covenants, it will never have the right to forfeit the lease in any circumstances.

Section 96 of the CLRA states that with the acquisition of the right to manage, management functions belonging to the landlord under each lease become instead functions of the RTM company. The same applies to any management functions belonging to any other party to the lease having management functions. Section 96(5) defines management functions to be functions relating to services, repairs, maintenance, improvements, insurance and management. But section 96(6) excludes from transfer any functions relating to re-entry or forfeiture. Section 97 states that any obligation owed by the RTM company to a leaseholder is also owed to the landlord (effectively putting the RTM company under a double legal duty).

Section 97(2) states that once management is transferred, the existing landlord or other manager party is no longer entitled to do anything which the RTM company is required or empowered to do under any lease except in accordance with an agreement made between the landlord and the RTM company. But this does not prevent the ground landlord from insuring the whole or any part of the premises at his own expense.

Sections 98 and 99 deal with the situation where a landlord's consent is required for anything which a leaseholder seeks to do in relation to the residential units. Chapter 4 has already explained the general legal principles which apply when a leaseholder seeks consent for any assignment, underletting, alterations or change of use. Sections 98 and 99 adapt these provisions to the situation where such responsibilities are taken over by the RTM company. In that situation, application for any landlord's consent must be made in the first instance by the leaseholder to the RTM company and not the ground landlord. However if the RTM company is minded to grant any approval on behalf of the landlord, Section 98(4) requires it first to give notice for that application to the ground landlord. At least 30 days notice to the landlord must be given where the approval sought relates to an assignment, under-letting, charging, parting with possession, making of structural alterations or improvements or change of use. This is reduced to 14 days where any other type of approval is sought. However no notice need be given to the ground landlord if the RTM company is minded to refuse the application — in which case it will be up to the RTM company to defend its decision if challenged.

If the ground landlord objects to the RTM company giving its consent to any leaseholder (and persists with that objection), section 99 only allows the RTM company to issue consent if sanctioned by an LVT. A similar situation applies if the ground landlord insists on imposing conditions on any approval.

A ground landlord who objects to the grant of any approval must give written notice of the objection (or any condition or requirement which must be satisfied) to both the RTM company and the leaseholder who has sought such consent. An application to an LVT may then be made by the RTM company, the leaseholder, any sub-tenant affected or the landlord.

Section 100 of the CLRA deals with the enforcement of leaseholder obligations. Section 100 refers to untransferred tenant covenants — meaning those covenants which do not relate directly to services, repairs, maintenance, improvements, insurance and management. The enforceability of untransferred tenant covenants may be exercised both by the RTM company as well as the landlord. Although as stated, only the ground landlord can exercise re-entry or forfeiture. The right which most leases reserve for ground landlords to enter individual residential units to check compliance with leaseholder obligations may be exercised by either the ground landlord or the RTM company.

Section 101 puts RTM companies under a general obligation to keep under review whether leaseholder covenants are being complied with and to report any non-compliance to the ground landlord. But the RTM company need not report any infringement by a leaseholder if:

- the failure has been remedied
- reasonable compensation has been paid in respect of the failure or
- the landlord has notified the RTM company that it need not report failures of a particular description.

Section 103 requires ground landlords to contribute to service charges in relation to any units in the building which are not held by a qualifying tenant (see above). These are known as excluded units. If the service charge proportions payable by the qualifying tenants do not add up to 100% of relevant service charge costs, the ground landlord must make up the difference in relation to the excluded unit(s).

To protect its right to manage, section 104 requires the RTM company to notify HM Land Registry and ensure that the appropriate entry is placed on the landlord's title (if registered). This is to alert anyone dealing with the landlord's title to the fact that the landlord's powers of management have been transferred to an RTM company. If this formality is not attended to, the rights acquired by the RTM company may not be binding on anyone dealing with the landlord's title.

Section 105 of the CLRA states that an RTM company's right to manage premises will cease (and revert to a landlord or other manager named in the lease), if agreed between the parties or if the RTM company is wound up, struck off the Companies Register or if creditors appoint a receiver. Its powers of management will also cease if a manager is appointed under Part 2 Landlord and Tenant Act 1987 (see above) — in which case the management functions will be transferred to the manager appointed by the LVT.

Section 106 prohibits any agreement associated with a lease which purports to exclude, restrict or otherwise obstruct the right of anyone to become a member of an RTM company or do anything associated with an RTM company.

Section 107 gives the county court jurisdiction to compel compliance with obligations relating to the exercise of the right to manage.

However no application can be made to the county court unless notice has been previously been given to the defaulting party requiring him to make good the default and more than 14 days had elapsed since the giving of that notice without his having done so.

The Meaning of Collective Enfranchisement

The statutory right to enfranchise means the right of leaseholders to collectively buy out the freehold of their block together with any intermediate leasehold interests standing between individual occupational leases and the freehold. But collective enfranchisement does not mean that each individual leasehold flat is converted into a freehold flat.

The occupational leases would remain and the terms and conditions contained in them would remain unchanged. But instead of an external ground landlord, the leaseholders will collectively become their own landlord enabling important management decisions to be made by the participating leaseholders on a democratic basis. The right to enfranchise differs from the right to manage in at least two respects namely:

- *It is a more expensive option* — as leaseholders will not only be taking over management responsibilities but will buy-out the landlord's interest. In fact most disputes about freehold enfranchisement are not about the principle of enfranchisement but how much the leaseholders should collectively pay the landlord for the freehold and any intermediate leasehold interests. There are complex statutory formulae for calculating this.
- *The leaseholders can exercise the full range of a ground landlord's powers and responsibilities.* The exercise of such powers and responsibilities are not restricted by law, as with right to manage. It follows that the leaseholders' management company will be able to extend individual leases.

Freehold enfranchisement can take place entirely by agreement, whereby the leaseholders (or some of them) will negotiate a price with the landlord for the purchase of his interest. But even where enfranchisement takes place by agreement, statutory procedures may still have to be followed to validate the transaction, particularly if not all the qualifying leaseholders are party to the negotiated acquisition. But it is more likely that leaseholders wishing to buy out a landlord's interest will prefer to adopt one of the several statutory procedures which have evolved since the Leasehold Reform Act 1967.

It was the 1967 Act (see Chapter 11), which first gave the lease-holders of houses (but not flats or maisonettes) the right to buy out their landlord's interest and become the freeholder of their own homes. Twenty years later Part I of the Landlord and Tenant Act 1987 gave the leaseholders of flats and maisonettes first right of refusal if their ground-landlord intended to sell its interest to a third party. But it was not until the Leasehold Reform Housing and Urban Development Act 1993 that the leaseholders of flats and maisonettes were first given a general right to acquire a landlord's interest at full market value so long as complex qualifying criteria could first be satisfied. That qualifying criteria was progressively simplified by the Housing Act 1996 and latterly the CLRA, so that long residential leaseholders now have a general right to enfranchise so long as a majority of them are willing and able to participate in the acquisition. There is a further right for leaseholders to compulsorily acquire a landlord's interest under part III of the 1987 Act in circumstances where landlord default can be proved or where a manager has already been appointed by the court (or LVT) for more than three years. With the exception of the right of first refusal (in which the purchase price is fixed by the landlord's notice), all the other statutory rights to enfranchise require professional valuation of the landlords interest utilising the relevant statutory criteria. Each of these statutory rights to enfranchise is dealt with in the following chapters.

The Right of First Refusal

Section 1 of the Landlord and Tenant Act 1987 requires any landlord intending to sell (or grant an intermediate lease out of) his interest in a block of flats to first give statutory notice in relation to that pending disposal, to the qualifying tenants of the flats contained in those premises.

Those provisions apply to the whole or part of a building containing two or more flats held by qualifying tenants so long as the number of flats held by such tenants exceeds 50% of the total number of flats contained in the premises. However the requirement does not apply where more than 50% of the internal floor area of the premises (in aggregate) is occupied or intended for occupation otherwise than for residential purposes. Therefore the requirement would not apply to flats above shops where the internal floor area of the shops below exceeds that of the flats above.

Section 2 makes clear that the duty to give notice only applies where it is the *immediate* landlord which is intending to sell its interest. Therefore if the immediate landlord's interest is itself held on lease from the freeholder, there will be no requirement on the freeholder to give notice before transferring the freehold interest to a third party. The rare exception is where the leasehold interest of the intermediate landlord is for less than seven years or which can be terminated within seven years, in which case it will be the superior landlord on whom the obligation lies.

Section 3 defines a qualifying tenant as being any residential leaseholder other than an assured or assured-shorthold or protected shorthold tenancy or business or certain agricultural tenancies.

Section 4 defines a "relevant disposal" of a landlord's interest to mean any type of disposal other than the lease of a single flat or any of those transactions listed in section 4(2). These excepted transactions include mortgages of the landlord's interest and other transactions of a close family nature, property transfers made on divorce or where the landlord becomes bankrupt.

Section 4A of the 1987 Act (as amended by section 89(1) of the Housing Act 1996) extends the obligation of the landlord to serve notice on leaseholders of a pending disposal to a third party before the landlord can even exchange contracts with that third party. Without section 4A it would have been possible for landlords to exchange contracts immediately with a third party conditional on compliance with the relevant notice requirements and on the leaseholders not exercising their right of first refusal. With section 4A, landlords cannot even enter into a conditional commitment to sell to a third party unless and until they have served notice on the leaseholders and given leaseholders the statutory opportunity to exercise their collective right.

Offer notices

Sections 5 to 5E of the Act (as amended) deal with the type of prior notice (known as an offer notice) to be served by the landlord on each of the leaseholders. There are five different types of notice depending on the manner in which the landlord's interest is to be marketed. Note in particular section 5(3) which prevents a landlord from "packaging" several buildings together in a single transaction. In those circumstances section 5(3) requires the landlord to deal with each building separately in complying with the statutory requirements.

Deputy High Court Judge Geoffrey Vos QC interpreted the effect of section 5(3) in the Chancery Division case of *Long Acre Securities Ltd* v *Karet* (2004) 11 EG138, which comprised an estate comprising approximately four separate structures with an access way, car parking areas, a central yard or forecourt, paths, roadways and an amenity space. The estate was known as the Frognal Estate and was held by landlords Long Acre on an underlease. The estate also comprised 55 residential flats held by the occupants on subunderleases. Long Acre proposed to sell its underlease by public auction and serve a single offer notice on each of the qualifying leaseholders relating to the whole of its estate. The leaseholders challenged the validity of the offer notice. The issue was whether the landlord had to serve a separate notice in respect of each of the four separate structures. Surprisingly Judge Vos ruled that they did not.

He said that it was necessary to give effect to the purpose of the legislation, which was to give leaseholders the right to acquire their landlord's interest and to provide a workable procedure to achieve that purpose. In the particular circumstances of this case, to try to split the transaction up would have made the legislation unworkable.

Section 5A describes the offer notice to be served when a landlord is proposing to sell its interest by traditional exchange of contracts followed by completion of the sale. In that case the notice must:

- detail the principal terms of the sale including the description of the property, the estate or interest in the property to be sold and the principal terms of the proposed contract (including the deposit and price required)
- state that the notice constitutes an offer by the landlord to the leaseholders to enter into a contract on those terms which may be accepted by the majority of qualifying tenants of the constituent flats
- provide a period of at least two months from the date of service of the notices during which the leaseholders may accept the landlord's offer
- provide a further period of two months during which the leaseholders may choose a nominee to acquire the landlord's interest on their behalf (if they have previously notified the landlord of their acceptance of his offer).

Section 5B deals with the offer notice to be served when a landlord intends to sell its interest by auction. That notice must set out the principal auction terms and state that the interest is proposed to be sold at public auction. It must also state that the notice constitutes an offer by the landlord, which may be accepted by the required majority of the qualifying leaseholders and specify a period of at least two months for the leaseholders to accept that offer. If the leaseholders decide to accept, that notice must also specify a further period of 28 days within which to choose their nominated purchaser. The notice itself must be served between four and six months before the date of the auction and the period during which the offer may be accepted must expire not less than two months before the auction. The period for choosing the nominee must also end not less than 28 days before the auction date. Unless the time and place of the auction and the name of the auctioneers are stated in the Offer Notice, the landlord must also provide the leaseholders with a further Notice (at least 28 days beforehand) providing the date, time and location of the auction.

A section 5C offer notice will apply where a landlord proposes to grant to a third party an option or right of pre-emption (that is to say right of first refusal) affecting his interest in the property. That notice must provide particulars of the principal terms of the proposed transaction describing the property, what the landlord is proposing to grant, the price and other principal terms in which the option or right of pre-emption will be exercisable, including the price when payable. The notice must state that it constitutes an offer by the landlord to grant *an option or right of pre-emption* in the same terms which may be accepted by the required majority of the qualifying tenants. As before the notice must also specify a two month period for acceptance and a further two month period for the leaseholders to choose their nominated purchaser.

Section 5D applies where the landlord proposes to move straight to a transfer of his interest without a prior exchange of contracts, option or right of pre-emption. As before, the notice must provide details of the proposed transaction, the price required, and the fact that it constitutes an offer by the landlord to sell to the leaseholders on similar terms, which can be accepted within the period of not less than two months — with a further two month period for choosing a nominee.

Section 5E deals with a rarer situation where a landlord proposes to sell his interest to a third party for something other than a monetary payment (for example, an exchange of other land). As well as providing details of the proposed transaction and the rights of the required number of qualifying leaseholders to accept the landlords offer and choose a nominee, a section 5E notice also requires leaseholders to be notified of their right to make an election under section 8C of the Act (as amended) as it might be impossible for the leaseholders to provide the non-monetary price required by the landlord for the transfer of its interest. Section 8C(4) enables the leaseholders to convert the non-monetary price to its equivalent financial value which (in the case of the example referred to) will be the value of the property to be exchanged. This monetary value will be assessed by an LVT if not agreed.

Acceptance notices

Once any of the five possible offer notices have been served, Section 6 of the Act prevents the landlord from selling or contracting to sell its interest to anyone save a leaseholders' nominee until after the initial deadline for service by the leaseholders of their acceptance notice. If an acceptance notice is served by leaseholders, the landlord is then prevented from selling its interest by a further statutory period during which the leaseholders may choose their nominee.

If a landlord has served an offer notice and the required majority of the qualifying tenants either do not respond with an acceptance notice within the specified period or — if having served an acceptance notice — do not notify the landlord of their chosen nominee within such further specified period, Section 7 of the Act allows the landlord 12 months within which to sell the specified interest to *anyone* but not on terms more beneficial to the purchaser than as previously offered to the leaseholders.

Where the offer notice was served pursuant to section 5B, the landlords interest can only be sold at public auction on the terms specified in the offer notice.

Section 8 deals with the situation where the required majority of qualifying leaseholders respond to the landlords offer notice with a valid acceptance notice followed up by notice of their chosen nominee purchaser. The landlord cannot then sell his interest except to the leaseholders' stated nominee. However at this stage, neither the landlord nor the nominee are under any binding obligation either to sell or proceed with the purchase. Within one month from receipt of the lessees' notice of nomination, section 8 (3) gives the landlord one month in which to notify the leaseholder's nominee that the landlord does not intend to proceed with the sale of the landlord's interest. If such notice is not served on the nominated person within one month, the landlord still has rights to withdraw from the transaction (see below) but must follow the correct statutory procedures.

Exchanging contracts

In the absence of a notice of withdrawal by the landlord under section 8(3)(a) above, Section 8A requires the landlord to issue a draft contract to the stated nominee for transfer of the landlord's interest to that nominee on the terms stated. Any failure to provide a draft contract within that period may be treated by the nominee as if the landlord had withdrawn formally from the transaction.

However if a draft contract is sent by the landlord to the nominee, that nominee has only two months in which to sign and exchange contracts and pay a 10% deposit or serve notice on the landlord indicating his intention no longer to proceed with the acquisition of the landlord's interest. Any failure by the nominee to respond to the contract within the two-month period will be treated as a withdrawal by the nominee from the transaction. If however the nominee is ready willing and able to exchange contracts within the two-month period but the landlord fails to complete the exchange within the following seven days, it will be treated as if the landlord had withdrawn from the transaction.

Auction sales

Section 8B applies to the situation where an offer notice has been served under section 5B in relation to a pending auction sale and where that auction of the property takes place and is subject to a successful bid. Section 8B(2) states that the leaseholders' nominee may, not less than 28 days before the date of the auction, elect that the provisions of this section shall apply. If such notice has been given and within seven days from the date of the auction, the landlord must send a copy of the memorandum of sale to the nominee. Within 28 days from receipt of that sale memorandum the nominee may serve notice on the landlord accepting its terms and fulfilling any other conditions (such as payment of the customary 10% deposit) to be fulfilled by the successful bidder on entering into the contract. If this is done, the sale memorandum will take effect as if the leaseholders' nominee and not the highest auction-bidder was party to the auction contract. The contractual completion time will then run from the date of service by the nominee of his notice under section 8B(4) and completion will take place otherwise in accordance with the sale conditions, so long as the nominee is given at least 28 days to complete the purchase. If the nominee either does not give notice under section 8B(2) electing that the provisions of that section apply or fails to serve notice accepting the terms of the contract and fulfilling any associated conditions, it will be treated as if the nominee had withdrawn from the transaction.

Section 8E deals with the situation where a landlord requires the consent of a third party (such as a superior landlord) before transferring his interest to the nominee and, in that event, requires the landlord to use his best endeavours to procure such consent and to issue court proceedings if it appears that such consent is being withheld unreasonably. However, if such consent cannot be obtained, the landlord may serve notice to this effect on the nominee in which case the landlord may dispose of his interest elsewhere but on no more favourable terms to the purchaser. In the event that the landlord is unable to proceed with the transaction due to the inability of obtaining the required consent, section 8E(6) enables the landlord to recover his expenses incurred in connection with the disposal, to a limited extent.

Notices of withdrawal

Until contracts are actually exchanged between the landlord and the leaseholders' nominee for the sale of the landlord's interest, either side can serve a notice on the other withdrawing from the transaction. Where it is the leaseholders' nominee who serves notice of withdrawal, the landlord will be free to sell his interest to any third party, but not on terms more favourable than those previously offered to the leaseholders. Where notice of withdrawal is served more than four weeks from the start of the nomination period specified in the offer notice, the party serving Notice will be responsible for any costs reasonably incurred by the other party from the expiration of those four weeks until the date notice of withdrawal is received.

Section 9A of the 1987 Act (as amended) enables the leaseholders' nominee to serve notice on the landlord indicating his intention no longer to proceed with the acquisition of the landlord's interest. The nominee is obliged to serve notice of withdrawal if he becomes aware that the number of the qualifying tenants desiring to proceed with the acquisition of the landlord's interest is less than the required majority of qualifying tenants of those flats. Where the nominee gives notice of the withdrawal, the landlord has twelve months from receipt of the notice to dispose of his interest to any third party. However, where the offer notice was made pursuant to section 5B (sale by auction), the sale can only be made at a public auction and otherwise on comparable terms to those specified in the offer notice. If notice of withdrawal is served within four weeks from the start of the nomination period, neither the nominee nor the qualifying tenants would be liable for any costs incurred by the landlord in connection with the sale of his interest. However, as stated, landlord's costs must be reimbursed if the notice of withdrawal is served later, and that liability will attach to each of the qualifying tenants as well as the nominee. The right to withdraw will not apply after contracts have been exchanged between the landlord and the nominee for the transfer of the landlord's interest.

Section 9B allows a landlord to serve notice of withdrawal on the nominee indicating his intention no longer to proceed with the sale of his interests. Where notice of withdrawal is given by the landlord, he is prevented from selling his interest for one year from the date of service of the notice of withdrawal. As with a nominee's withdrawal notice, the landlord is not liable for any costs incurred by the nominee if his notice is served within four weeks from the start of the nomination period. If the notice is served later, the landlord will be responsible for the nominee's costs from the expiry of the four week period until the actual receipt of the notice.

Section 10 applies to the situation where premises cease to be premises to which Part 1 of the 1987 Act applies, after the landlord has served the appropriate offer notice. In those circumstances the landlord can serve notice on the qualifying tenants stating that the premises have ceased to be premises to which this part of the Act applies and that the offer notice and anything done in pursuance of it is to be disregarded. Such notice must be served on at lease 90% of the qualifying leaseholders or where such qualifying tenants number less than ten, it must be served on all but one of them.

Landlord's failure to follow statutory procedures

Until the statutory provisions were strengthened by the Housing Act 1996, there was little leaseholders could do if a landlord failed to offer his interest to the leaseholders before selling on to a third party. If the leaseholders were aware of their rights and responded quickly enough, it was possible to insist on a retransfer from the purchaser of the landlord's interest back to the qualifying tenants. However in most cases, the third party acquired the landlord's interest without challenge.

Section 10A of the 1987 (as amended) now makes it a criminal offence for a landlord to transfer his interest (or part of it) to a third party, without having first served an offer notice under section 5 or having otherwise contravened any of the prohibitions or restrictions on the transfer of his interest. A landlord convicted of this offence can be fined up to £5,000.

Section 11 states that the rights of qualifying leaseholders are enforceable against the purchaser of the landlord's interest in circumstances where no offer notice has been served (see above) or if the transfer was made in contravention of any other statutory restriction arising out of these provisions.

Section 11A enables the required majority of qualifying tenants to serve notice on the purchaser requiring him to:

- to give particulars of the terms on which the sale was made (including the deposit and price paid) and the date on which the landlords interest was sold and
- where the transaction consisted of entering into a contract, to provide a copy of the contract.

Notice served under section 11A must specify the name and address of the person to whom (on behalf of the leaseholders) the particulars are to be given, or to whom a copy of the contract provided. There is also a four month time-limit for the service of such notice starting on the date notices under section 3A of the Landlord and Tenant Act 1985 (as amended) or documents of any other description indicating that the transaction is taking place and alerting the leaseholders to the existence of their statutory rights and time limits have been served on the required majority of qualifying tenants of the constituent flats. Section 3(A) applies where a new landlord is required to notify a qualifying leaseholder of his acquisition of the landlord's interest in circumstances where part I of the 1987 Act applied to the transaction. That notice is then required to tell the tenant that the disposal was one to which Part I of the 1987 Act applied; that qualifying tenants may have the right to obtain information about the disposal and to acquire the landlord's interest; and the time within which that right must be exercised. An incoming landlord who fails, without reasonable excuse, to give such notice, commits a criminal offence and can be fined.

The person served with a notice under section 11A must comply with it within a period of one month from the date for which it is served on him.

Where the landlord has exchanged contracts for the sale of his interest to a third party, but the transaction has not been completed, section 12A allows the required majority of qualifying tenants to serve notice on the landlord electing that the contract shall have effect as if entered into not with the contractual purchaser but with the leaseholders' nominee. Any such notice must be served within six months from either of the following dates:

- if a notice was served on the purchaser by the leaseholders under

section 11A (tight to information as to terms of sale), with the date on which the purchasers complied with that notice
- in any other case, with the date by which documents indicating that the original transaction has taken place and alerting the lease-holders to the existence of their statutory rights and time limits have been served on the required majority of the qualifying tenants.

However the leaseholders' nominee cannot take over any contract under section 12A unless he pays any contractual deposit and fulfils any other conditions required on exchange of contracts. Notwith-standing the terms of the contract, the leaseholders nominee cannot be required to complete the purchase before the end of 28 days from the date on which he took over the contract.

Section 12B applies either where completion of the transaction has taken place following an exchange of contracts or where the transaction has proceeded straight to completion without a prior exchange of contracts. The required majority of qualifying tenants may then serve a purchase notice on the purchaser requiring him to retransfer to the nominee the subject-matter of the original sale, on the terms on which it was originally made (including the price paid). A purchase notice must be served by the leaseholders within six months from the following dates:

- if notice was served on the purchaser under section 11A above (right to information as to transaction-terms), with the date on which the purchaser complied with that notice
- in any other case, with the date by which:
  1. notices under section 3A of the Landlord and Tenant Act 1985 (duty of new landlord to inform tenants of rights) were served relating to the original disposal or
  2. where section 3A does not apply, other documents indicating that the transaction has taken place and alerting the tenants to their statutory rights and relevant time limits have been served on the required majority of qualifying tenants.

Where the original transaction related to other property in addition to the building to which the statutory provisions apply, a purchase notice must require the purchaser only to transfer the building occupied by the qualifying tenants and to split the transaction terms to enable this to be done. In such a case the purchase notice may specify the subject matter of the original transaction and the terms on which the transfer to the nominee is to be made or may provide for those matters to be determined by an LVT. In circumstances where the purchaser has mortgaged the transferred interest, the retransfer to the nominee will operate to release the mortgage. Where the purchaser has carried out any other transaction in relation to the acquired interest, the price payable by the nominee to the purchaser will be reduced by an amount corresponding to any depreciation in the value of the transferred interest arising from that other transaction. Likewise, where the purchaser has done anything which has increased the value of the acquired interest, that will also be reflected in the price payable.

Section 12C applies to the rarer situation where an intermediate landlord has surrendered his intermediate lease to a superior landlord — effectively making the qualifying leaseholders direct tenants of the superior landlord. In those circumstances the qualifying leaseholders have six months in which to serve notice on the superior landlord requiring him to grant a new lease to the leaseholders' nominee on terms equivalent to the surrendered intermediate lease. The leaseholders' nominee must also reimburse any price paid by the superior landlord to the intermediate lessor in return for the surrender of the intermediate lease. Where the surrendered lease related also to premises other than those occupied by the qualifying tenants, the qualifying tenants may require the lessor to grant an intermediate lease which only relates to the relevant part with other terms of the original transaction split accordingly. As before the purchase notice may either specify the subject-matter of the transaction and the terms on which it is to be made or may provide for those matters to be determined by an LVT.

Miscellaneous

Section 12D allows for replacement of the leaseholders' nominee by another nominee only if he is no longer able to act as a nominee. Where there are two or more nominees and any of them ceases to act without being replaced, the remaining nominee or nominees may continue to act.

Section 13 gives LVTs general jurisdiction to hear and determine any question arising out of a notice served under sections 12A, 12B or 12C above and as to the terms of any split transaction where the original sale to a third party included more than one property.

Section 14 states that where a purchase notice has been served under sections 12B or 12C above, the leaseholders' nominee may at any time before a binding contract is exchanged, serve notice on the original purchaser (a notice of withdrawal) stating that he no longer intends to proceed with the transaction. Such a notice of withdrawal must be served if the nominee becomes aware that the number of qualifying tenants wishing to proceed with the transaction is less than the required majority. However if notice of withdrawal is served under section 14, the original purchaser may recover from the nominee (and any of the qualifying leaseholders) any costs reasonably incurred by him in connection with the transaction down to the receipt of the notice of withdrawal. Where such Notice of Withdrawal is served while proceedings are pending before a court or Lands Tribunal, the liability of the nominee for the purchaser's costs shall be as determined by the court or the tribunal. But the purchaser cannot recover costs incurred in connection with an application to an LVT.

Section 16 deals with a situation where the purchaser of the landlord's interest has already sold on to another party. In that situation, the purchaser shall, in response to any notice requesting information under section 11A, state the name and address of the person to whom he has sold on and also provide the end purchaser with the notice served under section 11A and the particulars given by him in response to it. Where a purchase notice has been served by the leaseholders' nominee under sections 12A, 12B or 12C, the purchaser shall immediately forward the notice to the end-purchaser and provide the nominee with written notice of the name and address of that end purchaser. Once the intermediate purchaser notifies the nominee and has forwarded the notice to the end purchaser, the statutory provisions shall instead of applying to the intermediate purchaser, apply to the end purchaser as if he were the transferee under the original transaction.

If at any time after notice has been served under sections 11A, 12A, 12B or 12C, the premises affected by the original transaction cease to be premises to which the statutory requirements apply, the purchaser may serve notice on the qualifying tenants stating that to be the case, in which case any notice served on him and anything done in response to that notice will be disregarded.

Of more importance to leaseholders is the three month time-limit for exchange of contracts set out in section 17(3). If contracts are not exchanged between the purchaser and the leaseholders' nominee within three months from service of the purchase notice and if there has been no application in connection with that notice with a court or LVT, the purchaser may serve on the nominee a notice stating that the purchase notice and anything done in pursuance of it will no longer have effect. Where an application has been made to an LVT or the court in relation to a purchase notice, the time-limit for exchange of contracts is two months from the outcome of those proceedings.

Note that although an LVT will resolve any questions arising out of a purchase notice, it is only the county court which can compel compliance with the statutory provisions. Section 19 states that the court may, on the application of anyone interested, make an order requiring any person in default of compliance with their statutory duty, to make good that default within such time as the court specifies. But an application cannot be made to the court unless notice has previously been served on the relevant person requiring him to make good the default and more than 14 days have elapsed since the service of that notice without him having done so.

Section 18A defines the required majority of qualifying tenants who need to participate to exercise the collective right of first refusal. There is one vote for each qualifying flat. The required majority is *more* than 50% of the available votes. Thus in a building with only two flats, both leaseholders must participate.

Protection of purchasers

Where it appears to a prospective purchaser of a landlord's interest that the transaction might be one to which part 1 Landlord and Tenant Act 1987 applies, the prospective transferee may take it upon himself to serve notices under section 18 on the leaseholders of the flats affected. But such notice must:

- inform each leaseholder of the general terms of the proposed transaction including the property to which it relates and the price to be paid
- invite the leaseholder to serve Notice on the prospective purchaser stating
    1. whether the landlord has already served an Offer Notice under section 5 of the Act and
    2. if the landlord has not served an offer notice, whether the leaseholder is aware of any reason why he is not entitled to be served with any such notice by the landlord and
    3. if the leaseholder is not so aware, whether he would wish to avail himself of the right of first refusal conferred by such notice if it were served and

- inform the leaseholder of the effect of section 18.

Where the prospective purchaser has served notice on at least 80% of the qualifying tenants and either:

- not more than 50% of the recipients have responded constructively within two months or
- more than 50% of the recipients have served notices on the prospective purchaser indicating that they either do not regard themselves as being entitled to receive an offer notice under section 5 or would not wish to avail themselves of that right of first refusal if such a notice were served, the premises affected by the transaction shall be treated as if the right of first refusal did not apply to it.

Drafting of notices

It is recommended that any notices to be served by landlords or leaseholders under part I of the 1987 Act be professionally drafted. This is because the legislation does not provide any pro forma notices. A notice which is incorrectly drafted or which omits vital information may be legally ineffective and statutory rights might be lost. The need for meticulous drafting was underlined in the case of *Tudor* v *M25 Group Ltd* [2004] 1 EGLR 23 when a leaseholders' notice requiring information about a recent transaction under section 11A of the Act (see above) was unsuccessfully challenged on the grounds that it did not contain the addresses of the participating qualifying leaseholders as required by section 54(2) of the Act.

Section 54 states that any notice required or authorised to be served under the 1987 Act must be in writing and may be sent by post. Furthermore any notice purporting to be served by the required majority of qualifying tenants must specify the names of all the persons by whom the notice was served and the addresses of the flats of which they are qualifying tenants.

The defendants M25 Group Ltd acquired the freehold of Brambridge House in Hampshire on 3 December 2001. A section 5 notice had been served by the previous freeholders on the qualifying tenants but there was a dispute as to whether the subsequent sale of the freehold on 3 December 2001 was compliant with Part I of the Act. It was not until 13 May 2002 that the leaseholders were notified of the change of landlord.

On 27 May 2002 solicitors acting for the leaseholders served statutory notice on M25 under section 11A of the Act requiring information about the recent transaction. M25 did not respond. However their solicitors wrote to leaseholders' solicitors on 29 May 2002 questioning the validity of their statutory notice. On 8 September 2002 the leaseholders gave notice under section 19 (see above) requiring compliance. The statutory 14 days elapsed without response and the leaseholders applied to the county court for an order compelling compliance.

The landlord disputed the proceedings saying that the leaseholders' section 11A notice was invalid because it did not state the addresses of all the named leaseholders as required by section 54(2). If that notice was invalid, it followed that all the leaseholders rights under the 1987 Act to acquire the freehold were lost. The sole issue for the Court of Appeal was whether the failure to state the leaseholders' addresses was fatal to the validity of the section 11A notice. In spite of the omission, the Court of Appeal ruled that the notice was a legally valid request for information under section 11A.

The purpose of section 11A was to enable leaseholders to seek information about recent transactions and for landlords to know who sought the information and whether they equalled a majority of the qualifying tenants who were entitled to that information. One had to question the effect of the omission and whether it mattered. The important thing was that the parties were in an existing landlord and tenant relationship, which M25 had undertaken voluntarily. The Act proceeded on the assumption that certain facts were readily and indisputably ascertainable and it was permissible to construe section 54 in that light. M25 would not have had any practical difficulty in obtaining the leaseholders' addresses. Practical considerations entitled the court to adopt a flexible approach and the court was entitled to decide the consequences of non-compliance with a technical requirement in the particular statutory context in which it arose. The substantive provisions of the Act conferred on the leaseholders the right to acquire the freehold. The secondary provisions included the notice requirements of section 11A and the formal requirements of section 54, including the address requirements, which were merely supportive. Their omission did not invalidate the section 11A notice.

The General Right to Enfranchise/Extend Under the 1993 Act

Part 1 of the Leasehold Reform, Housing and Urban Development Act 1993 gives residential leaseholders of flats and maisonettes a general right to enfranchise (that is to say, buy out a freehold interest) provided basic qualifying criteria are met. Those criteria are:

- that at least two thirds of the flats in the building are occupied on long residential leases
- that leaseholders representing at least one half of the *total* number of flats in the building are willing to participate in the purchase of the freehold
- that not more than 25% of the building is occupied for non-residential purposes
- that between them the participating leaseholders can afford the anticipated capital cost of buying out the freehold interest and any intermediate leasehold interests standing between the occupational leases and the freehold.

Before the CLRA took effect, there were other qualifying hurdles to be overcome, which have since been abolished. Essentially all that a residential leaseholder now has to demonstrate to be counted is that they hold an occupational lease which was originally granted for more than 21 years. No longer do leaseholders have to prove that they have personally been in actual occupation of the particular residential unit. Therefore there is no reason why company leaseholders cannot participate in the same way as private individuals.

Taking on board the reforms introduced by the CLRA (some of which at the time of writing have yet to take effect), the procedure resembles that of acquiring the right to manage in that it involves the setting up of a special purpose company; service of a participation notice on other leaseholders; and a notice/counternotice procedure for exercising the right, with any disputes being resolved through the county court or LVT. But the procedure is substantially more complex and, if successful, will result in a conveyancing transaction. It is therefore recommended that professional expertise be sought from the start both as to the implementation of the correct legal procedures and as to valuation of the freehold interest and any intermediate leasehold interests. A more detailed overview of the procedure is set out below.

Right to enfranchise companies

Provided the basic qualifying criteria can be met, the first formal step for any group of leaseholders wishing to buy out the freehold interest, is to set up a right to enfranchise, or RTE, company to represent the participating leaseholders and to act as the nominee purchaser of the freehold interest.

At the time of writing, the government has not yet published regulations governing the constitution of RTE companies, although the basic requirements are set out in sections 4A to 4C of the 1993 Act (as amended by CLRA). Section 4A states that an RTE company must be a private company limited by guarantee whose memorandum of association states that its objects, or one of its objects, is the exercise of the right to collective enfranchisement of the relevant premises. Subject to appropriate adjustments in its name and constitution, it would seem that an existing RTM company (see Chapter 6) which has already taken over management of the building could convert itself into an RTE company for the purpose of buying out the freehold.

Section 4B of the Act restricts membership of an RTE company to the qualifying tenants of flats contained in the building and (if the company is already an RTM company) any landlords who currently have membership of that company. But note that membership of those landlords will cease as soon as the freehold and any intermediate leaseholds are finally transferred to the RTE company, by which time the landlords will not have any interest in the premises.

Who is the landlord?

In most cases there will only be one ground landlord with whom leaseholders have to deal: the freeholder. But in rare cases there may be a series of intermediate leases standing between the freehold and the occupational leases of individual flats within the building. For example freeholder A may have granted a 999-year lease of land to developer B for the construction of a block of 20 flats. B then builds the block of flats and grants a 99-year underlease to each flat buyer. Therefore, to exercise their right to enfranchise, the participating occupational leaseholders must buy out not only freeholder A but also developer B, to take full control of their own affairs. Section 9(2) of the Act deals with the situation and defines the particular landlord (or reversioner) with whom the RTE company is to deal when exercising its right to enfranchise. It does this by reference to part 1 of schedule 1 of the Act which states that the reversioner in this situation will be the freeholder unless, by agreement between the landlords, the court appoints someone else. Section 9(2A) (as amended) deals with the even rarer situation where ownership of the freehold is split between more than one person and requires the identity of the relevant reversioner to be determined in accordance with part 1A of schedule 1 to the Act.

The resident landlord exception

Some buildings will not qualify for enfranchisement under the 1993 Act where the ground landlord is in residential occupation of part of the building and can satisfy the criteria set out in section 10 of the Act (as amended).

This exception will never apply to a purpose built block of flats but may apply to converted premises where the same ground-landlord has owned the freehold of the building since before the conversion took place and he, or another adult member of his family, have occupied a residential unit within the premises as their only or principal home throughout the period of twelve months ending with the claim to enfranchise. Under section 10(5) an adult family member is described as a spouse, child, parent, or in-law of any of the above.

Right to obtain information

In most cases, full information about the interest of the freeholder as well as that of any intermediate leaseholder, as well as the interests of other occupational leaseholders within the block, can be obtained direct from the appropriate Land Registry District Office. However there are many older titles which are still unregistered, particularly as regards freeholds, or leases which have remained in the same ownership for many years. In such circumstances the leaseholders will have to make enquiries direct to their immediate landlord as well as any superior landlord or freeholder (if different) to obtain the information they need to exercise their right.

Section 11 of the Act gives *any* qualifying leaseholder the right to give written notice to his immediate landlord (or to anyone collecting rent as the landlord's agent) requiring the recipient to give the leaseholder (so far as is known to the recipient) the name and address of every freeholder as well as the name and address of anyone holding an intermediate lease standing between the immediate ground landlord's interest and that of the freeholder. The qualifying leaseholder may also give written notice directly to the freeholder (if different from the immediate ground landlord) requiring the freeholder to give the name and address of everyone known to him (other than the qualifying leaseholder himself) who holds a lease either of the whole of the building (an intermediate lease) or any known tenant or licensee of all other parts of the building. The leaseholder making the request also has the right to seek other information reasonably required by him to participate in the right to collective enfranchisement of the freehold. Section 11(5) also gives any qualifying leaseholder the right (on giving reasonable notice) to be provided with a list of any document held by the landlord or freeholder which would be reasonably required by an RTE company in making an enfranchisement claim and which on a voluntary sale of the landlord's (or freeholder's) interest would be expected to have been made available to the prospective buyer of the landlord's interest. Section 11(5) also gives qualifying leaseholders the right to inspect any of the listed documents at any reasonable time and on giving reasonable notice and a further right (on payment of a reasonable fee) to be provided with a copy of any such documents. Anyone served with a request for information under section 11 must provide that information within 28 days.

The participation notice

After having been incorporated but before taking any other formal step towards the acquisition of the freehold, the newly established RTE company must first serve a notice inviting participation under section 12A of the 1993 Act (as amended by the CLRA). This notice must be given to every qualifying tenant of those flats in the building who had not already agreed to become a participating member of the RTE company. A notice of participation to participate must:

- state that the RTE company intends to exercise a statutory right to buy out a freehold
- state the names of those members of the RTE company who are already participating
- explain the rights and obligations of members of the company with respect to the exercise of the right (including their joint obligation to meet the price payable in respect of the freehold and any other interests to be bought in and associated costs),
- include an estimate of that price and those costs and
- invite the recipients of the notice to become participating members of the RTE company.

As with the invitation to participate in an RTM company (see Chapter 6), the invitation notice for an RTE company must also be accompanied by a copy of the memorandum of association and articles of association of that company or state where those documents can be inspected and copied.

Note that even at this early stage of the process, the participating leaseholders should already know approximately how much the freeholder (as well as any intermediate leaseholders) would be entitled to charge for their interests. Therefore before serving a participation notice on other leaseholders, a professional and reasoned valuation of the landlord's interests should have already been carried out. The general principles governing valuation are set out below.

Claiming the right (the initial notice)

Subject to compliance with the preparatory steps outlined above and provided that at least one half of all the flats in the building are owned by qualifying leaseholders who have taken up active membership of the RTE company, the company is in a position to service its initial notice on the ground landlord claiming its right to collective enfranchisement of the freehold. Section 13 of the 1993 Act sets out the detailed requirements for an initial notice, which must:

- be served on the ground landlord (or where the ground landlord is not the freeholder, on the relevant person ascertained in accordance with schedule 1 as the reversioner)
- be given by or on behalf of the participating leaseholders comprising not less than one half the total number of flats in the building
- specify and be accompanied by an accurate plan clearly showing the whole of the premises to which the initial notice relates
- contain a statement of the grounds on which it is claimed that the right to enfranchise applies
- specify any intermediate leasehold interest which also has to be acquired to complete the enfranchisement
- specify the proposed purchase price for the freehold (and separately for any intermediate leasehold)
- state the full names of *all* the qualifying tenants of flats contained in the building together with the addresses of their flats and providing sufficient details of each of their leases, including the date the lease was entered into, the term of the lease and when that term commenced
- specify a date (at least two months ahead) within which the recipient of the notice must provide a constructive response.

Once given, the initial notice will remain in force until it is either replaced by a binding contract for sale (or court order) for the transfer of the freehold (and any intermediate interests) to the RTE company — or that a notice is formally withdrawn or is deemed to be withdrawn as a matter of law (see below).

It will be seen that the drafting of an initial notice is itself a complex technical process. As well as the valuations of the landlord's interest obtained before service of the participation notice, there is now the need to commission professionally drawn plans to accompany the initial notice. As soon as the notice has been served, it will also need to be registered against the landlord's title to ensure that any third party proposing to acquire any interest in that title is alerted to the existence of the initial notice and bound by it.

Landlord's response to initial notice

On receipt of an initial notice the ground landlord (or other recipient of the notice) is given by section 17 a statutory right of access to every part of the building in which he has a freehold (or intermediate leasehold) interest, where he needs such access to enable him to value his own interest in the building to which the initial notice relates. The RTE company also has the statutory right of access to any part of the building specified in the initial notice. In either case, the party seeking access must attend at a reasonable time and give at least 10 days prior notification to the occupier of those premises.

Section 18 places the RTE company under a general duty to disclose to the ground landlord (or other reversioner) the existence of any agreement between the RTE company and a third party under which the landlord' interest (once acquired) will be sold on to that third party. However this duty to disclose subsales does not apply to any mortgage negotiated by the RTE company to facilitate the purchase.

Section 20 states that the ground landlord may, within 21 days, require the RTE company to prove the title of any participating member of the company to the lease of which they claim to be a qualifying tenant. The RTE company must provide a constructive response within 21 days to the landlord's notice, failing which the initial notice will automatically lapse if the absence of that qualifying leaseholder would reduce the number of participating leaseholders to below the 50% threshold.

Section 21 deals with the reversioner's counternotice — which constitutes the landlord's formal response to the leaseholders' initial notice. The landlord who does not serve a valid counternotice on the RTE company within the two month deadline for responding may afterwards have difficulty in challenging the assertions made in the initial notice (including possibly the leaseholders' valuation of the freehold interest). Like the initial notice, the landlord's counternotice should be drawn up professionally and served after taking appropriate valuation advice. The counternotice not only requires the landlord to state whether he admits or disputes the basic right to enfranchise but also provides the only opportunity by which the landlord can dispute any of the terms offered by the participating leaseholders (including price) and make counter proposals. The initial notice and the landlord's counternotice are the most critical documents in the process as they are the documents which set out each party's stall and provide the starting point for subsequent negotiation or dispute resolution.

Section 21 states that the reversioner's counternotice must contain one of the following three statements:

- that the reversioner admits that the RTE company was on the relevant date entitled to exercise the right to collective enfranchisement
- that, for reasons specified, the reversioner does *not* admit that the RTE company was so entitled or
- either admit or dispute the RTE company's entitlement but at the same time state the landlord's intention to redevelop the whole or a substantial part of the premises (see below).

If the counternotice admits (in principle) the right of the RTE company to enfranchise, the reversioner's notice must then state which of the leaseholders' proposals contained in the initial notice are accepted by the reversioner and which (if any) are not. In relation to those leaseholder proposals which are disputed, the reversioner must then make his counter proposals in the counternotice. The reversioner must also state any proposals for a lease back to him of any parts of the building following acquisition to the RTE company (see below).

The counternotice must also state on behalf of the landlord:

- which additional landlord interests the reversioner requires the RTE company to acquire, on the grounds that those additional interests would for all practical purposes cease to be of use or benefit to the landlord or could not be reasonably managed or maintained by the landlord
- which rights (if any) the landlord wishes to retain in respect of the building on the grounds that those rights are necessary for the proper management or maintenance of other property in which he retains a freehold or leasehold interest and
- any other provisions which the reversioner or other relevant landlord considers should be included in any transfer to the RTE company. The reversioner may also include proposals for the leaseback to him of certain flats and other units in the building which are not occupied by qualifying leaseholders.

Where, in his counternotice, a reversioner disputes the right of the RTE company to buy out the freehold and any intermediate leasehold interests, section 22 gives the company power to apply to the county court for a declaration that the company has established its statutory right to enfranchise. But any application to the court by the RTE company must be lodged within two months from within receipt of the counternotice, failing which the initial notice will lapse. If the court upholds the RTE company's claim to enfranchise, the court will give the reversioner an opportunity to give a further counternotice to the RTE company responding in detail to the proposals contained in the initial notice. The one exception is where the counternotice has been served on the grounds that the landlord intends to redevelop and an application by the lessor for an order under section 23(1) (see below) is pending or the period for making an application for such order has not expired. If the RTE company's application under section 22 is dismissed by the court, the initial notice ceases to have effect at that date.

A reversioner who has given a counternotice stating that he intends to redevelop the building can apply to the court under section 23 for an order declaring that the leaseholders' right to enfranchisement shall not be exercisable in relation to the building because of the landlord's stated intention to redevelop the whole or a substantial part of the building. However section 23(2) prevents the court making such an order in the landlord's favour unless it is satisfied:

- that not less than two thirds of the long leases of flats within the building are due to terminate within five years and
- that for the purposes of redevelopment, the landlord intends, once those leases have terminated
  (i)   to demolish or reconstruct or
  (ii) to carry out substantial works of construction on the whole or a substantial part of the premises and
- that he could not reasonably do so without obtaining possession of the flats to which those leases relate.

Any application for an order under section 23 must be made by the landlord within two months from the giving of his counternotice to the RTE company. If the court finds in the landlord's favour, the initial notice will cease to have effect. Where the landlord's claim under section 23 is rejected by the court, the court must give the landlord an opportunity to give a further counternotice responding to the proposals made in the initial notice.

Section 25 deals with the situation where a reversioner fails to give a counternotice (or further counternotice) within the two-month deadline specified in the initial notice. Section 25(1) then gives the RTE company the right to apply to the court to determine the terms on which it is to acquire the freehold and any intermediate leasehold interest. Those terms ordered by the court may include provision for the leasing back to the landlord of flats or other units not held by qualifying tenants (see above). However, before making such an order, the court must still be satisfied that the RTE company was on the relevant date entitled to exercise its right of collective enfranchisement in relation to the building and that it has complied with the relevant procedural requirements. Section 25(4) gives RTE companies only six months from the counternotice (or further counternotice) deadline in which to make its application to the court.

The following case should be noted in relation to reversioners counternotices.

*   *9 Cornwell Crescent London Ltd* v *Kensington and Chelsea London Borough Council* [2005] 14 EG 128 — when the Court of Appeal ruled that the landlord's counternotice was not invalidated by the fact that the figure quoted by the landlord for the sale of its interest was unrealistically high. In that case the leaseholders had quoted a proposed purchase price of £210 against the landlord's quoted sale price of £130,000, based on a professional valuation. The fact that the landlord's counternotice was held to be valid meant that any outstanding dispute over the purchase price would have to be resolved through the LVT. The appeal judges said that the county court should be wary of attempting to resolve disagreements as to valuation at the notice and counternotice stage by developing a form of "strike out" or different procedure. Lack of good faith in specifying the sale price was the only ground on which the validity of the landlords counternotice could be challenged.

Dispute resolution — sale terms

Once a landlord's counternotice (or further counternotice) has been given admitting the RTE company's right to acquire the freehold and any intermediate leasehold interests, the parties have a further two months in which to resolve by negotiation any differences as to the amount of the purchase price or other terms of sale. Where any issues remain outstanding at the end of that two month period, either party may apply to the LVT to determine such matters as are in dispute. But no application can be made to the LVT more than six months after receipt of the landlord's counternotice (or further counternotice). Where all the sale terms have been either agreed between the parties or determined by an LVT but a binding contract has not been exchanged between the parties incorporating those terms within the period of two months from the date those terms were finally agreed or determined by the LVT, the RTE company may apply to the *court* providing for the landlord's interests in the freehold and any intermediate leaseholds be transferred to the RTE company on the specified terms. But note also that there is only a subsequent two month period for making such applications to the court.

Untraceable landlord

Chapter 2 of this book provides a practical example of freehold enfranchisement in circumstances where there has been no contact between leaseholders and their landlord for many years and attempts to trace the ground landlord had proved unsuccessful. In such circumstances the procedure to be adopted by the RTE company to obtain the freehold is to be found in sections 26 and 27 of the Act (as amended).

In those circumstances (and after reasonable efforts have been made unsuccessfully to trace the ground landlord) the RTE company can apply direct to the county court for an order transferring the freehold (or any relevant intermediate interests) to the RTE company on terms determined by an LVT, following the making of the "in principle" court order that the landlord's interest be transferred. However, before making an order transferring (or vesting) the landlord's interest to the RTE company, the court must first be satisfied that the applicants were, as a matter of law, entitled to exercise their right to enfranchise — in other words that the building satisfied the statutory requirements, there was a sufficient proportion of qualifying leaseholders participating in the purchase, and that an RTE company had been established and a participation notice served.

Before transferring the landlord's interest, the court may also require the RTE company to take further steps by way of advertisement or otherwise to trace the missing landlord. If the landlord is traced as a result of those further efforts, the proceedings will cease and then the rights and obligations of the parties will be determined as if an initial notice has been given exercising the right — with the newly discovered landlord being given an opportunity to serve a counternotice.

Section 27 deals with the role of the LVT in relation to the transfer from an untraceable landlord. If all efforts to trace the missing landlord fail, the RTE company will be required to pay the purchase price into court in return for the transfer. A former landlord who later reappears will then collect what is owing to him from the court but will have no other claim against the leaseholders or RTE company.

The case of *R (on the application of Ford (T/A David Sayer)) v Leasehold Valuation Tribunal*, LTL 2 March 2005 compares the respective roles of the county court and the LVT in an untraceable landlord situation.

As a result of a Land Registry mistake, a former garage converted into living accommodation, which was separate from the flats and retained by the landlord, had been included within the title of one of the qualifying leases. Having been unable to contact the ground landlord, the three qualifying leaseholders set up a special purpose company to acquire the freehold and applied to Wandsworth County Court for an order vesting title to the whole of the freehold, including the converted garage, in the special purpose company. That order was granted in the landlord's absence.

The leaseholders then made their application to the LVT to determine the price payable, which was assessed by valuing the former garage as a garage and not as living accommodation. The relevant sum was paid into court by the leaseholders and title was transferred. The leaseholders then took the first step to obtain physical possession of the garage by affixing a notice to it.

A sharp eyed acquaintance of the ground landlord spotted that notice and alerted that landlord, Mr Ford (who called himself Sayer) — who then returned to England to try to salvage the situation and get his converted garage released from the transfer to the special purpose company. He first tried unsuccessfully to overturn the county court decision transferring the freehold. He then turned his attention to the LVT but unfortunately could not appeal to the Lands Tribunal as (being an absent landlord) he was not technically party to the original LVT proceedings. Instead he succeeded in overturning the entire decision of the LVT by applying to the High Court for judicial review on the grounds that the members of the LVT had misled themselves as to how they could have dealt with an obvious discrepancy. In granting judicial review, Collins J said that an LVT had discretion under section 27 of the Act to go behind the terms of a vesting order previously obtained through the county court and decide that not all of the landlord's freehold title should be transferred to the lessees' special purpose company. But he acknowledged that there were limitations on what could be expected from the LVT. It was not there to provide representation for the absent landlord nor was it under any duty to seek out evidence not already before it or evidence which ought have been before it.

If the evidence actually produced before the tribunal disclosed significant discrepancies, the tribunal ought to make further enquiries to ascertain the true position and ensure that no injustice was done to either party. In this case there was an obvious and apparent conflict between one of the qualifying leases and the freehold for the property. The tribunal ought therefore to have contacted the Land Registry to investigate the matter further and its failure to do so was an error of law. The end result was that the matter had to be considered afresh before a differently constituted LVT.

Withdrawal of initial notice

Section 28 of the Act (as amended) allows the RTE company to withdraw an initial notice by giving a notice of withdrawal to a relevant landlord at any time before a binding contract is exchanged between the RTE company and the landlord for the acquisition of the landlord's interest. However if such a notice of withdrawal is issued, the RTE company and every leaseholder participating in it will be jointly liable for payment of the landlord's abortive costs from the date of the receipt of initial notice until receipt of the notice of withdrawal. Those costs will include not only the landlord's legal costs but also the costs of valuing his interest and other administrative expenses.

The importance of observing deadlines is underlined by section 29 of the Act which deems the initial notice to be withdrawn if the RTE company does not make its application to the court (or as the case may be, the LVT) within the stated time-limits. Again there will be similar cost consequences for the participating members of the RTE company. The other consequence of withdrawal or deemed withdrawal of the initial notice is that no further notice can be served within 12 months from the date of the withdrawal.

Price and valuation

Fixing the price to be paid by the leaseholders for the purchase of the landlord's interest is (save for the simplest cases) the most complex part of the enfranchisement procedure. It is also the question most likely to give rise to dispute between the lessor and qualifying leaseholders. The actual price payable by the leaseholders is to be determined in accordance with section 32 of and schedule 6 to the Act.

In the most common situation where a group of qualifying

leaseholders are dealing with a single landlord who owns the whole of the freehold without any intervening leases, article 2 of schedule 6 states the price payable to be the aggregate of:

- the value of the freeholder's interest — that is to say the amount which the lessor might be able to achieve if he sold his interest to a third party on the open market
- the freeholder's share of marriage value — meaning the "added value" which arises when a freehold and the leases derived out of it are all brought into the same collective ownership, instead of being owned separately and
- any amount of compensation payable to the freeholder to compensate for any reduction in value of the interest of the freeholder in other property resulting from the enfranchisement or other loss or damage referable to his ownership of that other property.

Broadly the value of a ground landlord's interest equates to what it might be worth to a prospective buyer as an investment. This will be determined, among other things, by the lengths of the leases still outstanding and the amount of each ground rent and whether it is fixed or reviewable. The longer the lease and the smaller the ground rent, the less leaseholders will have to pay their landlord to buy out his interest. But as those leases shorten, the value of the landlord's interests will rise. The starting point for any group of leaseholders wishing to buy out the landlord's interest is, "what is it worth to the landlord in terms of investment?" Although leaseholders may be tempted to fix their own ball park figure before embarking on the enfranchisement process, at some stage in the process the leaseholders will be required to justify their valuation in accordance with recognised valuation techniques. This is why a professional valuation is so essential before any formal step is taken in the process.

Article 4 of schedule 6 states that marriage value is to be split 50:50 between the outgoing ground-landlord and the acquiring RTE company. However marriage value is to be disregarded entirely in relation to any lease which has more than 80 years unexpired. Article 4(2) further defines marriage value to be that increase in value attributable to the potential ability of the participating members of the RTE company to have new leases granted to them without payment of any premium or restriction as to length of term.

When there are separate intermediate leasehold interests as well as the freehold, the purchase price for each of those intermediate leasehold interests will be separately assessed but will not include any marriage value. Note that as well as the purchase price the leaseholders (acting through the RTE company) will also have to pay the landlord's costs in relation to the enfranchisement which will comprise:

- any investigation reasonably undertaken as to whether particular leaseholders qualify or as to any other question arising out of the initial notice
- proving the landlord's title
- valuing the landlord's interests
- the actual costs of conveying the landlord's interest.

However, a landlord's costs are only regarded as "reasonable" to the extent that they might reasonably be expected to have been incurred by the landlord if he had to pay for them himself without reimbursement.

Transfer of landlord's interest

Save in the case of an untraceable landlord, normal conveyancing procedures will apply once contracts have been exchanged between the RTE company and the landlord for the purchase of the landlord's freehold (or intermediate leasehold interest). Section 35 states that the transfer to an RTE company pursuant to statutory procedures will be freed from any mortgage affecting the landlord's title. Simultaneously with the purchase or the freehold interest, there may be requirement for the RTE company to lease back relevant parts of the building to the landlord pursuant to schedule 9 of the Act (see above).

The actual conveyancing process is set out in schedule 1 of the Leasehold Reform (Collective Enfranchisement and Lease Renewal) Regulations 1993 (as amended). Its main provisions are:

- that once the leaseholders' claim has been established, the RTE company may require the landlord to prove his documentary title
- once evidence of the landlord's title has been provided, the company has 14 days to raise any queries
- the reversioner must provide a draft contract to the company within 21 days of settlement of the acquisition term: the company then has 14 days in which to propose any amendments
- the company must pay the usual 10% deposit on exchange of contracts (or £500 if greater).

Extending a lease

Although commonly referred to as a lease extension, what Chapter II of the Leasehold Reform Housing and Urban Development Act 1993 does is to give individual leaseholders the right to a longer replacement lease, which will completely supersede the existing lease. By this means the lease term may be extended by 90 years from the contractual expiry of the existing lease term.

Although the procedure for extending a lease loosely follows the procedure outlined in Chapter I of the Act for buying out a landlord's freehold, the main difference is that it is not a collective right. So long as the basic qualifying criteria are met, any individual leaseholder can apply to their landlord for a lease extension. There are also special statutory formulae governing how much each individual leaseholder must pay for their extension.

As with freehold enfranchisement, the statutory hurdles which a leaseholder must overcome to qualify for an extended lease have been relaxed by the CLRA. Section 39 of the 1993 Act (as amended) states the remaining qualifying criteria to be:

- that the tenant holds a ground lease originally granted for a term of more than 21 years and
- that the tenant for the previous two years has been a qualifying tenant of the flat.

There is no requirement for the leaseholder claiming the extended lease to be in actual occupation of the flat to which the application relates. Even if leaseholders die, their personal representatives will have two years from the grant of representation to claim an extended lease.

Section 40 specifies who is the landlord, to whom any claim for an extended lease is to be addressed. The relevant landlord is:

- the immediate landlord, if that landlord holds either the freehold or an intermediate lease which is long enough to enable the extended underlease to be granted out of it or
- (where the immediate landlord is not the freeholder or holder of a long enough intermediate lease), the freeholder or any intermediate lessee (if any) whose interest is sufficient to enable the extended lease to be granted out of it. Where this scenario applies, that superior landlord will conduct on behalf of all the other landlords, all proceedings arising out of the leaseholders'

claim for an extended term. In that situation schedule 11 of the Act requires the leaseholder to serve copy his notice claiming the right to on everyone who is known to have a landlord's interest.

Section 41 enables any qualifying leaseholder to give to his immediate landlord or landlord's agent a notice requiring the recipient to state whether that landlord holds the freehold and, if not, to give the leaseholder such information that is known to the immediate landlord about:

- the name and address of the freeholder
- the length of the immediate landlord's intermediate-leasehold interest (if applicable) and the extent of the premises in which it subsists and
- the name and address of anyone who has a leasehold interest superior to that of the immediate landlord, the duration of such interest and the extent of the premises to which it relates.

Where the immediate landlord is not the freeholder, the tenant may give to the freeholder a notice requiring him to give the tenant such information as is mentioned above (so far as is known to the freeholder). The tenant may also give notice to anyone holding a leasehold interest superior to that of the intermediate landlord. Any notice given by the tenant to anyone under section 41 shall also require the recipient to state whether he has received notice under section 13 (see above) seeking freehold enfranchisement and if so to provide a copy of such notice. Section 41(6) provides a 28 day deadline for any recipient to respond to the tenant's notice seeking information.

Notice claiming the right

Section 42 deals with the notice to be served by the qualifying leaseholder on the relevant landlord (see above) claiming the right to an extended lease. That notice must be given not only to the relevant landlord but also to any other organisation which is already party to the ground lease (such as a management company). Section 42(3) deals with the content of the tenant's notice which must:

- state the leaseholder's full name and the address of the flat in respective of which he is claiming a new lease
- contain the following particulars, namely

    (1) sufficient particulars of the flat to identify it

    (2) sufficient particulars of the tenant's lease including the date on which it was signed, the term for which it was granted and when the term commenced.

- specify the premium which the tenant proposes to pay for the grant of the replacement lease and where any other sum is payable by him (see below), the amount which he proposes to pay in accordance with that provision

- specify the terms which the tenant proposes should be contained in any such lease (see below)

- state the name of anyone representing the tenant in connection with his claim together with an address in England and Wales to which notices may be given to him in relation to his claim

- specify the date by which the landlord must respond to the notice by giving counternotice under section 45 (see below).

Personal representatives acting for a deceased leaseholder have two years from the grant of probate or letters of administration to lodge their notice claiming an extending lease. Section 42(5) requires the tenant's notice to specify a date not less than two months ahead within which the landlord is to respond. Note that once notice has been served by the tenant under section 42, no further notice may be served in relation to the same flat while that notice remains in force. Note also that if that notice is subsequently withdrawn or lapses (see below), the tenant will lose the right to serve a further notice within a period of 12 months immediately following the withdrawal.

Section 43 provides that once a tenant's notice has been served claiming the right to an extended lease, that notice will bind any third party taking over the landlord's interest (whether by purchase or otherwise) as if a contract had been entered into between the landlord and the tenant. An outgoing tenant may also assign the benefit of any current notice to anyone taking over the tenant's existing lease.

As with an initial notice claiming freehold enfranchisement, it is recommended that any notice claiming an extended lease be professionally drawn by the person who will be carrying out the eventual conveyancing transaction if the notice succeeds. Likewise a professional valuation (utilising the valuation principles and formulae set out in the Act) should also be obtained before the notice is served. The amount of this valuation will then constitute the leaseholder's financial offer for the extended lease and may have to be defended in any future LVT proceedings, in event of dispute.

Landlord's response

Once notice has been served claiming the right to an extended lease, section 44 gives the landlord (and any landlord's professional representative) a right of access to the flat to enable the landlord to make his own valuation. That right of entry is exercisable at any reasonable time and on giving not less than three days notice to the tenant.

Section 45 deals with the landlord's counternotice, which must be served on the tenant within the critical deadline specified in the tenant's notice claiming the right. Like the tenant's notice, the landlord's notice should also be professionally drawn after the landlord has taken the appropriate valuation advice.

Section 45(2) states that the counternotice must:

- state that the landlord admits that the tenant had on the relevant date the right to acquire the new lease
- challenge the tenant's claims to an extended lease and specify the grounds of that challenge or
- contain a statement as mentioned under (a) or (b) above but contend that the landlord intends to challenge an extended lease on the grounds that he intends to redevelop any part of the building in which the flat is contained (see section 47).

If the counternotice admits (in principle) the tenant's right to a new lease, the counternotice must then state which of the proposals contained in the tenant's notice are accepted by the landlord and which of those proposals are disputed and in relation to each disputed proposal, the landlord must state his counter proposal.

Where the landlord gives a counternotice disputing the validity of the tenant's claim for a replacement lease, section 46 places the onus on the landlord to apply to the court for a declaration that the tenant was not entitled to a new lease. An application must be made within two months from the giving of the counternotice. If such application is not made within the two month deadline or having been made, is subsequently withdrawn, matters will proceed as if the landlord had not given any counternotice. However, if the landlord is successful in his application to the court, the tenant's notice claiming the right will cease to have effect.

Section 47 enables the landlord to defeat a tenant's claim for a replacement lease on the grounds that the landlord intends to redevelop any part of the building within which the relevant flat is contained.

Application must be made by the landlord to the court to that effect and section 47(2) prevents the court from making such an order unless satisfied that:

- the tenant's existing lease is due to terminate within five years
- that for the purposes of redevelopment the landlord intends, once the lease has so terminated, either to demolish or reconstruct the building, or carry out substantial works of reconstruction on the whole or a substantial part of the building in which the flat is situated, and that the landlord cannot reasonably do so without obtaining possession of the flat.

The landlord's application under section 47 must be made within two months from the giving of the counternotice to the tenant. The landlord can of course only apply under section 47 if those grounds have already been stated in the landlord's counternotice.

Dispute resolution relating to terms etc

As has already been seen, disputes over the in principle validity of the tenant's claim for an extended lease are to be dealt with through the county court. However, once the tenant's in principle right has been established, the detail over the premium and other terms and conditions of the new lease are to be resolved through an LVT if not agreed.

Section 48 states that where the landlord has given the tenant a counternotice admitting the right to a lease extension but any of the terms of acquisition remain in dispute at the end of the following period of two months, an LVT may on the application of either party, determine the matters in dispute. However such application must not be made later than six months beginning with the issue of the counternotice (or further counternotice where appropriate). Where all terms of acquisition have either been agreed between the parties or determined by an LVT but a replacement lease has not been signed within the following two months, the court may on the application of either party make such order as it thinks appropriate as regards the performance or discharge of any obligations arising out of the claim. A failure by the tenant to enter into a lease on the terms agreed might result in the lapse of the tenant's claim (section 48(4)) — in which case there will be a 12 month interval before the tenant can apply again.

In circumstances where the landlord has failed to give the tenant a counternotice within the specified deadline (or a further counternotice where appropriate), the tenant may then apply to the court for an order requiring the landlord to grant a replacement lease in accordance with the terms specified in the tenant's original notice claiming the right. But the court cannot make such an order unless satisfied that the tenant had the statutory right to acquire a replacement lease and that the tenant's notice was properly served in accordance with the requirements in schedule 11 of the Act. Section 49(3) also specifies a six month deadline for the tenant to make his application to the court, beginning with the deadline for receipt of the counternotice (or further counternotice as the case may be). Where the court has set out the terms on which the replacement lease is to be granted but the new lease has not been entered into pursuant to that order within a further two months (or other specified period), the court may make such other order as it deems appropriate with respect to the performance or discharge of any outstanding obligations. Where it is the tenant who has failed to sign the lease that order may also treat the tenant's claim as withdrawn.

Untraceable landlord

As with freehold enfranchisement, the court may make an order vesting title in the leaseholder. But in this case it is not the transfer of the landlord's freehold but the grant of a replacement lease out of that freehold. In those circumstances section 50 enables the tenant to apply direct to the county court for such a vesting order. Section 50(3) prevents the court from making such an order unless satisfied that the tenant had the right to acquire a replacement lease and that he was not precluded by any other statutory provision from giving a valid notice under section 42 (for example if a previous notice had been withdrawn within the previous 12 months).

Before making a vesting order, the court may require the tenant to take further steps by way of advertisement or otherwise to trace the missing landlord. If as a result of those further steps the landlord is traced, the matter will continue as if the tenant had given formal notice to the landlord claiming a new lease — giving the landlord an equivalent right to respond by counternotice.

Section 51 states that a vesting order (in those circumstances) is an order providing for the surrender of the tenant's existing lease in return for the grant to him of a new lease on such terms as may be determined by an LVT. On payment of the specified sum into court, the relevant district-judge (or other specified person) will execute a lease which is in a form approved by an LVT and which otherwise gives effect to the order.

Withdrawal or lapse of tenant's notice

Section 52 enables any tenant claiming an extended lease to withdraw that notice at any time before the lease is actually entered into. This might arise if the tenant subsequently discovers that he is no longer financially able to acquire the extended lease. However even if the notice is not actually withdrawn, statute will treat it as having been withdrawn if the tenant fails to comply with relevant time-limits for applying to the court under section 48 or section 49 (above).

Section 54 automatically suspends the operation of a tenant's notice claiming a replacement lease if there is a pending claim for freehold-enfranchisement (see above) under Chapter 1 of the Act. Where the tenant's right is suspended because of this, section 54(3) requires the landlord to give notice to the leaseholder of such suspension.

Completing the extended lease

Once a qualifying leaseholder has given notice of his claim for a replacement lease under section 42 of the Act (see above) (and if necessary substantiating that claim) section 56 of the Act requires the landlord to grant to the tenant a new extended lease in replacement for the tenant's existing lease. That new lease will be for a term of 90 years from the contractual expiry date of the existing date. Thus if an existing lease has 25 years to run and a replacement lease is granted under Chapter II of the 1993 Act (as amended), that replacement lease will be for a term of 115 years in aggregate. Whatever ground rent is paid under the existing lease, under the new lease that ground rent will reduce to zero. However the leaseholder will be required to pay a premium to the landlord in return for the replacement lease, which will be calculated in accordance with the principles laid down in schedule 13 of the Act. This premium will comprise the aggregate of the following:

- the diminution in value of the landlord's interest in the tenant's flat
- the landlord's share of the marriage value (see below) and
- any compensation payable to the landlord under paragraph 5 of

schedule 11 for any diminution in value of the landlords interest in any other property resulting from the grant to the tenant of the replacement lease.

When both landlord and leaseholder commission their respective valuations in relation to the tenant's claim for an extended lease, care must be taken to ensure that the valuation is carried out strictly in accordance with the principles of schedule 11.

"Diminution in the value of the landlord's interest" is not the same as the "open market value" of the extended lease and may enable the leaseholder to purchase the extended lease on more beneficial terms than if an equivalent extended lease is acquired on the open market. Paragraph 3 of schedule 11 sets out how the diminution in the value of the landlord's interest is to be calculated and the assumptions, which must be made regarding this. Like freehold enfranchisements, the split marriage value between landlord and tenant is 50:50 and this is again defined in schedule 11. A new subparagraph 4(2A) requires marriage value to be disregarded where the unexpired term of the tenant's existing lease exceeds 80 years.

As well as the premium the leaseholder must also discharge any other financial obligations owing to the landlord under the existing lease as well as the landlord's costs in relation to the extended lease. If the exact amount of any of these sums has not yet been calculated (such as pending service charges), the leaseholder must provide the landlord with reasonable financial security for such payment.

Save as to the peppercorn ground rent and the extended length of the replacement lease, section 57 requires it to be granted in other respects on the terms of the existing lease with such modifications as may be necessary to deal with any alterations to the property made since the grant of the original lease. Section 57(2) makes clear that the landlord is entitled to impose adequate service charge provisions in the new lease to cover expenses incurred by him, whether or not such provisions were included in the original lease. However clause 57(6) preserves the right of the landlord and leaseholder to agree any variations required in the new lease to remedy any defect in the existing lease or which it might be reasonable to do to bring the document up to date. Where there is a management company or other person or organisation joined as a third party to the existing lease, they are also required to be made party to the new lease, so long as that third party is not placed under any obligation subsisting beyond the contractual expiry of the existing lease. Section 57(11) requires the new lease to include a statement compliant with current land registration rules, which state that it is a lease granted pursuant to section 56 of the Act.The important thing to note about section 56 is that the new lease is in substitution to the existing lease, which will cease to have effect.

The Leasehold Reform (Collective Enfranchisement and Lease Renewal) Regulations 1993 (as amended) deal with the conveyancing formalities. These regulations have already been referred to in accordance with freehold enfranchisement. Schedule 2 of those regulations deal with extended (replacement) leases under Chapter 2 of the Act.

Regulation 2 enables the ground landlord to demand a 10% deposit of the amount proposed in the tenant's claim notice (or £250) whichever is greater. That deposit is to be paid to the landlord's solicitor or licensed conveyancer as stakeholder. However that deposit is returnable if the tenant's claim is withdrawn or otherwise lapses or ceases to have effect. Unlike freehold enfranchisement, there is no requirement for contracts to be exchanged and the transaction will proceed straight to the replacement lease.

Regulation 4 entitles the landlord to require that the leaseholder proves title to his existing lease and gives evidence by statutory declaration of his occupation of the flat. Notice requiring this must be given within 21 days of service of the tenant's claim notice.

Once the tenant's claim for a replacement lease has been established (either because the landlord has admitted the claim or the tenant has obtained a court order to that effect), regulation 5 obliges the landlord to prove documentary title to his interest in the property if required by the leaseholder.

Once the landlord's title has been disclosed to the tenant under regulation 5, regulation 6 gives the tenant 14 days in which to raise any objections or questions ("requisitions") on the landlord's title as disclosed. The landlord must then reply to those objections or requisitions within the following 14 days — giving the tenant a further seven days in which to comment on those answers.

Regulation 7 gives the landlord 14 days from agreement or determination of final terms to prepare a draft lease and deliver it to the tenant (or the tenant's lawyers). The tenant then has 14 days from the receipt of the draft lease to propose any amendments, which must be given in a statement. If the tenant does not give formal notice stating any required amendments, the tenant will then be required to accept the landlord's draft document as drawn. Once the format of the draft lease is agreed, the landlord's lawyers will then prepare the final version (as an original and counterpart — see above) for signature by both parties.

Once the draft lease has been approved, regulation 8 gives either the landlord or the leaseholder the right to give the other not less than 21 days requiring completion of the replacement lease.

Mortgages

As it is quite likely that either the tenant's existing lease or the landlord's freehold or intermediate leasehold interest (or probably both) is under mortgage, the position of mortgage lenders has to be accommodated within the legislation.

Section 58 makes clear that fact that the landlord's ownership of the freehold (or intermediate lease) is in mortgage does not prevent the leaseholder from claiming and acquiring an extended lease from the landlord. Any extended lease will therefore be binding on the mortgagee of the landlord's interest.

Section 58(4) deals with the situation where it is the tenant's existing lease which is subject to mortgage. The replacement lease will then take effect subject to the same mortgage in substitution for the existing lease. The mortgage lender will therefore be entitled to custody of the replacement lease together with the remaining deeds/documents following its grant.

Repeated renewal

Section 59 of the Act makes clear that acquiring an extended lease under chapter II of the Act does not prevent the leaseholder (or any transferee of the extended lease) claiming a further extended lease in the future. Thus it would seem that leaseholders could in theory repeatedly extend their leases indefinitely subject to payment of the required premium. However once the extended lease has expired (and in the absence of further extension) the leaseholder will not have any of the statutory security of tenure which benefits other leaseholders on the expiry of their leases.

Landlord's costs

Once a leaseholder has given notice to his landlord under section 40 claiming an extended lease, the leaseholder will be required to reimburse the landlord's reasonable costs of and incidental to any of the following matters, namely:

- any investigation reasonably carried out by the landlord into the tenant's right to an extended lease
- any valuation of the tenant's flat required for the landlord to assess the appropriate premium or other sums which the landlord is entitled to demand
- the landlord's conveyancing costs in relation to the grant of the new lease.

In assessing the reasonableness of the landlord's stated costs, section 60(2) limits them to such costs as a person might incur if he was paying them out of his own pocket.

Termination for redevelopment

Section 61 of the Act gives landlords a limited right to terminate an extended lease if the landlord intends either to demolish or reconstruct the building or to carry out substantial works or construction on the whole or a substantial part of the building within which the flat is contained and that the landlord could not reasonably do so without obtaining possession of the flat. In those circumstances the landlord must first apply to the court for an order that he is so entitled to repossess the flat on payment of compensation to the leaseholder for the termination of the extended lease. However an application for termination on this ground can only made within the following time-limits:

- during the period of 12 months ending with the date on which the original lease would have expired had it not been extended or
- at any time during the period of five years ending with the expiry of the replacement lease.

Where an extended lease is prematurely terminated under section 61, the amount of compensation payable to the leaseholder will be calculated in accordance with schedule 14 to the Act.

Compulsory Acquisition of a Landlord's Interest

Provisions enabling qualifying tenants to compulsorily buy out a landlord's interest bolt-on to a leaseholder's right to apply to the court to appoint a manager where an existing ground landlord is in serious breach of his obligations relating to repair, maintenance or management of a block of flats. Compulsory acquisition (or enfranchisement) under Part III of the 1987 Act differs from the leaseholders' general collective right to enfranchise under the 1993 Act (see Chapter 9) in the following respects:

- it only applies where a manager has already been appointed under Part II of the Act more than two years previously or following proof of landlord default
- application must be made to the county court
- it must be proved that the ground landlord remains in breach of his obligations under the leases and that the breach is likely to continue
- the court has a discretion whether to allow compulsory acquisition and will only do so if it considers this appropriate
- there is no requirement for the leaseholders to set up a special-purpose company to acquire the freehold: it can be transferred to anyone they nominate
- where the defaulting ground landlord is himself an intermediate lessee (and does not own the freehold), the procedure only allows the leaseholders to buy out the defaulting landlord's intermediate leasehold interest.

Qualifying criteria

1. The building must comprise two or more flats held by "qualifying tenants" — meaning residential lessees holding leases originally granted for more than 21 years (but qualifying status will not apply to any leaseholder owning more than two flats of the building).
2. The total number of flats held by qualifying tenants must be at least two thirds of the total number of flats comprised in the premises.
3. Compulsory acquisition cannot apply if more than 50% of the internal floor area of the premises is occupied otherwise than for residential purposes.
4. Compulsory acquisition cannot apply where the landlord is an exempt landlord or a resident landlord or where the premises are included within the functional land of any charity.

Procedure

Step one

Notice under section 27 of the Act must be served on the ground landlord by a majority of the qualifying tenants, which must:

*   specify the names of the qualifying tenants on whose behalf it is served, the addresses of their flats and the name and address of someone in England and Wales on whom the landlord may serve any notices
*   state that those tenants intend to apply to the court for an acquisition order but (if a breach is capable of remedy within a reasonable time) they will not do so if the landlord complies with specified requirements
*   specify the grounds on which they will ask the court to make an acquisition order and the matters on which they rely to establish those grounds
*   where matters are capable of being remedied by the landlord, the notice must require the landlord, within such reasonable period as the notice specifies, to take specified steps for the purposes of remedying the situations.

At the time of writing there is no prescribed format for a notice under section 27. However it is recommended that any such notice should be professionally drafted, as it will form the basis of any subsequent application to the court under this legislation. It follows that if the notice is defective, the entire application is likely to fail.

If the ground landlord is uncontactable, a preliminary application must be made to the county court under section 27(3) for an order dispensing with requirement to serve a notice. In granting such dispensation, the judge may direct that other notices are served or other steps taken.

Step 2

Apply to the county court for an acquisition order.

Unlike other enfranchisement procedures, there is no provision for a landlord's counternotice, and the leaseholders may move straight to the court if the ground landlord fails to comply with the requirements of the preliminary notice served under section 27 (if the landlord's breach can be remedied at all). That application to the court made under section 28 of the Act, and as with the preliminary notice, must be made on behalf of a majority of the qualifying tenants. Section 29 then enables the court to grant an acquisition order if:

- the landlord is in breach of an obligation owed by him to the lessees relating to the management of the premises, or (in the case of an obligation dependant on notice) would be in breach but for the fact that it would not be reasonably practicable for the tenant to give him the appropriate notice (eg where part of the building is in disrepair and the tenant has been unable to contact the landlord about it, and those circumstances are likely to continue) — or

- the court has previously appointed a manager under Part II of the Act (following proof of landlord default) more than two years previously and that appointment remains in force — or

- the court considers it appropriate to make an acquisition order in the circumstances of the case.

Any application to the court for an acquisition order under section 28 is of course speculative as there is no guarantee that an order will be granted even if the grounds are proved. An acquisition order is most likely to be granted where a manager has already been appointed and there is no realistic prospect of the ground landlord resuming its responsibilities in the foreseeable future. Provided the relevant qualifying criteria can be met, leaseholders may be better advised to use the more up to date procedures contained in Part I of the 1993 Act to buy out the freehold interest (see Chapter 9).

Section 30 sets out the content of an acquisition order (once granted), which is to provide for the leaseholder's nominee to acquire the landlord's interest on such terms as may be agreed between the landlord and the qualifying tenants or (in default of agreement) by an LVT following an application made under section 31. Note also section 30(2), which states that an acquisition order may be granted subject to such conditions as the court thinks fits and, in particular its operation may be suspended on terms fixed by the court. Suspension of an acquisition order will give a ground landlord one final opportunity to comply fully with his management responsibilities before his interest is transferred to the leaseholders' nominee.

Where the landlord's interest is itself leasehold (or subject to any mortgage or other third party interest) and the landlord cannot transfer his interest without the consent of a third party, section 30(5) requires the landlord to use his best endeavours to obtain that consent and (if necessary) institute court proceedings against a third party if it is apparent that consent is being unreasonably withheld.

Step 3

Implement the acquisition order.

The most important preliminary step is for the leaseholder to register the acquisition order under the Land Registration Act 2002 (or Land Charges Act 1972 if the landlord's title is unregistered), to ensure that any third party dealing with the landlord's interest is alerted to the existence of the acquisition order. In fact that application to registration should already have been made as a pending land action when proceedings are first instituted.

Where the terms of acquisition cannot be agreed between the parties, application may be made to an LVT under section 31 to determine the terms on which the landlord's interest is to be transferred to the leaseholders' nominee according to what appears to the LVT to be fair and reasonable.

Where there is disagreement as to the amount payable by lessees to the ground landlord for the acquisition of his interest, the LVT will assess this as an amount which, in their opinion, the landlord's interest might be expected to realise if sold on the open market (but on the fictitious assumption that none of the existing leaseholders were buying or seeking to buy that interest). That formula (which makes no mention of marriage value — see Chapter 9) suggests that leaseholders acquiring the landlord's interest under Part III of the 1987 Act (on proof of landlord default) may be able to do so at a cheaper price than leaseholders exercising their statutory right to enfranchise (without proof of fault) under the Leasehold Reform Housing and Urban Development Act 1993 (see Chapter 9). However, there are no landmark decisions on the subject.

Unless otherwise agreed or ordered by the court, section 32 of the 1987 Act states that any acquisition of the landlord's interest under Part III of the 1987 Act takes free from any mortgage of the landlord's interest.

Where a landlord cannot be found or his identity cannot be ascertained, section 33 allows the court to transfer the landlord's title to the leaseholder's nominee on terms defined by the court and on payment into court by the leaseholders of what the landlord would otherwise expect to be paid if contactable. Unlike other statutory enfranchisement, that amount will not be fixed by an LVT but by a surveyor selected by the President of the Lands Tribunal to certify his opinion of the amount for which the landlord's interest might be expected to realise if sold on the open market. As well as the acquisition price, the leaseholders must also pay into court the amounts or estimated amounts remaining due to the landlord from all leaseholders under the terms of their leases.

Discharge of acquisition order

Section 34 of the Act allows a landlord to apply to the court to discharge an acquisition order in any of the following circumstances:

- if the nominated person has had a reasonable time to acquire the landlord's interest in pursuance of the earlier order but has not done so or
- if the number of qualifying tenants (see above) desiring to proceed with the acquisition of the landlord's interest is less than the required majority of qualifying tenants of flats in that building or
- if the premises have ceased to be premises to which Part III of the 1987 Act applies.

In summary a ground landlord can seek cancellation of an acquisition order if he can prove unreasonable delay on the part of the leaseholders' nominee in completing the transaction or if there has been a material change of circumstances, which make the leaseholders no longer eligible to acquire the landlord's interest. This is most likely to arise if several of the leaseholders back out of the transaction once it is known how much they will have to contribute to the buying out of the landlord's interest.

Section 34(2) gives participating qualifying tenants the right to serve notice on the landlord indicating their intention no longer to proceed with the acquisition of the landlord's interest. Section 34(3) requires a leaseholders' nominee to give such notice to the landlord if he becomes aware that the number of qualifying tenants currently participating in the transaction is less than is the required majority or if the premises have ceased to be premises of which Part III of the Act applies.

Late withdrawal by qualifying leaseholders (or their nominee) from the acquisition of their landlord's interest also has an adverse cost implication for those leaseholders as they will be required to reimburse any costs reasonably incurred by the landlord in connection with the sale of his interest down to the time notice of withdrawal is received. Each of the qualifying leaseholders who participated in the proposed acquisition will be jointly liable for the whole of the landlord's costs — not just their individual proportions.

Freehold Enfranchisement of Houses

The Leasehold Reform Act 1967 is the earliest legislation giving leaseholders the right to either buy out their landlord's freehold or alternatively to extend their lease.

The legislation only applies to leasehold houses, not to flats or maisonettes. The Act sets out the basic structure for freehold enfranchisement and lease extension which was followed in the 1993 Act. But there are differences.

The Leasehold Reform Act 1967 is not collective — meaning that any residential lessee of a house can apply for enfranchisement so long as other qualifying criteria can be met. As with flats and maisonettes, the person claiming the right must hold a lease originally granted for more than 21 years. The former value and low rent tests no longer apply to freehold enfranchisement but still apply to lease extensions. An extended lease will not be granted in return for a capital premium but instead at an increased ground rent following the formulae laid down in the 1967 legislation.

As with the 1993 Act, enfranchisement under the 1967 Act follows a notice and counternotice procedure, with disputes being referred either to the court or to an LVT as appropriate.

Enfranchisement procedure

The procedure for the tenant to exercise the statutory right to buy out a landlord's interest or extend a lease is set out in section 5, schedules 1–3 and associated regulations made under the Act. These regulations are:

- The Leasehold Reform (Enfranchisement and Extension) Regulations 1967 — as amended from time to time (most recently by the Leasehold Reform (Enfranchisement and Extension) (Amendment) (England) Regulations 2003).
- The Leasehold Reform (Notices) Regulations 1997 as amended by the Leasehold Reform (Notices) (Amendment) (England) Regulations 2002.

Section 5 sets out general provisions relating to freehold enfranchisement or lease extension:

- That once a tenant gives notice of intention to exercise the right to enfranchise or extend, the rights and obligations arising from that notice will bind their executors or administrators (in case of death) and anyone to whom the benefit of the right is assigned as if a contract for sale or lease has been entered into between the parties which was binding on the estate.
- A tenant's rights and obligations from such a notice may also be assigned by the tenant to any third party taking over the lease of the house (but not assigned separately). If a lease is assigned without a contemperaneous assignment of the notice (and as a result the notice ceases to have effect) the former tenant is liable to compensate the landlord for any interference (if any) resulting from the notice affecting the landlord's power to dispose of or deal with his own interest in the house and premises or any neighbouring property. In a case of *South* v *Trustees of the Philimore Kensington Estate* 2001 (case reference LTL19/10/2001) — Lightman J ruled that a purported assignment of the right to enfranchise which (contrary to section 5) was made independently of or before an assignment of the actual lease to which it was related, had no legal effect whatsoever.
- Any default by either landlord or leaseholder in complying with their respected obligations arising from a notice to enfranchise may be sued as a breach of contract.
- The notice served in accordance with the provision of the Act to enfranchise or extend the lease is registerable against a landlord's title at HM Land Registry (or if that title is unregistered — as a land charge) as an estate contract, to alert anyone dealing with the landlords interests of the existence of the notice. However the fact that a leaseholder may qualify generally for a right to enfranchise

or extend the lease is not itself registerable until such a time as notice is served.

- Once a leaseholder's notice has been served under the Act, this will override any contract which the landlord has already entered into with a third party for the disposal (in any manner) of the landlord's interest in the house or any part thereof, unless that contract provides for the contingency of a possible tenant's notice.
- The fact that a tenant has already applied for an extended lease does not prevent a subsequent notice to enfranchise the freehold.

Section 6A preserves the right of the personal representatives of a deceased tenant (if the tenant had already acquired the right to enfranchise or extend before death) to give notice claiming the freehold or extended lease within two years from the grant of representation.

Once a valid notice has been given by the leaseholder under section 5 of the Act, section 8 obliges the landlord to transfer the freehold at the price and on the terms and conditions set out later in this chapter.

The procedure itself is set out in Part II (Procedural Provisions) of schedule 3 to the 1967 Act (as amended), which can be summarised as follows:

A tenant's notice exercising either the right to freehold enfranchisement or an extended lease will normally be that set out as Form 1 in the Leasehold Reform (Notices) Amendment (England) Regulations 2002). That notice is in three parts:

(1) the first part which formally makes the claim;
(2) the second part containing the "particulars supporting tenant's claim" — being the factual information about the tenant's lease on which the tenant relies in support of the claim and
(3) statutory notes to assist the person drafting the notice as well as the recipient of that notice. Dealing with the first part of the form:

Clause 2 requires leaseholders to state whether they are seeking the freehold or an extended lease.

Where there are no intermediate leases between the occupational lease and the freehold, clause 4 requires the landlord/freeholder to respond within two months with a counternotice (see below) stating whether or not the leaseholders' claim for enfranchisement or an extended lease is admitted by the landlord and, if not, stating the grounds on which the claim is disputed. If the immediate landlord is not the freeholder (because there are intermediate leases), whether the recipient of the notice is required to respond within two months depends on whether that recipient is the person described as the reversioner for the purposes of the claim.

Clause 6 requires the leaseholder to state whether there is anyone else on whom the same notice is also been served. Clause 7 requires the landlord to copy the received notice to anyone else who is known or believed to have an interest in the property which is superior to the landlord's interest and record on that copy the date on which the landlord received the notice. The name and address of any other person must then be added to the list at clause 6 and written notice given to the leaseholder claiming the right.

The particulars which the leaseholder needs to include in the second part of the notice to support the claim are:

- the address of the house
- particulars of the house and premises sufficient to identify the property to which the claim relates. Therefore express reference must be made to any garage, outhouse, garden, yard and anything else which at the relevant time is leased to the tenant with the house.

Clause 3 requires particulars of the relevant lease (or leases) comprising the tenancy to prove that the tenancy has at all times been a long tenancy or treated as such. Those particulars must contain the date, names of the original parties, commencement date of the lease, and term of the lease with an adequate description of the property to which it relates. As already stated, to qualify as a "long lease" it must be for a term of more than 21 years. However where there have been successive tenancies of the same property, these may be aggregated together to determine whether (in total) there is a long tenancy. In any event the leaseholder *must* provide details and particulars of each successive tenancy (where applicable). Where a lease has already been extended under the Act, the date of the extension and the original contractual expiry date of the lease should be given. Clause 4 requires the leaseholder to state the date on which he acquired the lease. This is to demonstrate compliance with requirement that the claimant has owned the lease for two years prior to the date of the application.

Paragraph 6 of the form requires the leaseholder to provide details of any other long tenancy of the house or of any flat forming part of the house held by anyone. This provision again relates to the case of intermediate leases in light of the situation that under section 1 (1ZA) of the Act (as recently amended by the CLRA) intermediate lessees do not have rights to enfranchise or extend the lease where there exists "inferior" (meaning leases ranking below and deriving out of the intermediate lease), where it is that inferior tenant who has the right to enfranchise or extend the lease under the Act. In those circumstances the intermediate lessee can only claim the right to enfranchise or extend the lease if he can satisfy a "residence requirement", namely that tenant has lived in the property as his only or main residence for the last two years or for periods totalling two years in the last ten years.

Where there is a flat situated within the house which is let to someone who is a qualifying tenant for the purposes of Part I Leasehold Reform Housing and Urban Development Act 1993 (see Chapter 9) or if any part of the claimant's tenancy is a business tenancy (see below) paragraph 7 requires the applicant to demonstrate compliance with the residence requirement by giving the following particulars:

- the periods for which in the last 10 years, and since acquiring the tenancy, the applicant has and has not occupied the house as his residence
- describing what parts (if any) of the house have not been in the applicant's own occupation, and for what periods and
- what other residence (if any) the applicant has had and for what periods, and which is the applicant's main residence.

Paragraph 8 requires additional particulars showing the value of the house to demonstrate that it does not exceed the applicable financial limit — save where such particulars are not required under statute.

Paragraph 9 requires additional particulars (where applicable) to show the basis of statutory valuation of the premises under the foregoing paragraph.

Paragraph 10 requires additional particulars where the applicant is a trustee, personal representative or family member succeeding to the tenancy on death — to demonstrate their entitlement to enfranchisement or an extended lease.

Validity of notices

As with other forms of enfranchisement, it is recommended that Notice 1 exercising the right to enfranchise or an extended lease under the 1967 Act is professionally drawn — as any material errors may be fatal. However salvation may be found in paragraph 6(3) of the third schedule to the Act which states that the notice shall not be invalidated by any inaccuracy in the stated particulars or any misdescription of the property to which the claim extends. Where the claim extends to property not properly included within the lease or does not cover all the property that should be included, the court may allow correction of the notice under such terms as it deems fit. But paragraph 6(3) does not excuse all errors and in particular will not apply where the information required by statute has been omitted as opposed to being merely erroneous. This is apparent from the following cases.

- *Byrnlea Property Investments Ltd* v *Ramsay* [1969] 2 QB 253, where a Part I Notice was held invalid because the claimant had failed to clarify whether he was seeking either the freehold or an extended lease. A notice in such an uncertain form could not give rise to the statutory contract provided by section 5 of the Act.
- *Cresswell* v *Duke of Westminster* [1985] 2 EGLR 151. In providing information about his occupation of a property during the last 10 years, the leaseholder mistakenly failed to disclose that he had lived in Paris for six months. Nor did he disclose that he had a second residence, a cottage owned by his mother, although it was never suggested that this was his main residence. In that case the Court of Appeal upheld the decision of the trial judge that those inaccuracies were rescued by paragraph 6(3) of the third schedule and did not invalidate the notice.
- *Speedwell Estates Ltd* v *Delziel* (2002) 1 EGLR 55 — when the Court of Appeal were presented with notices containing a series of defects on which they had to adjudicate. Each of these defects and the court's view of them are summarised below:
  (1) Existing leases were incorrectly and incompletely described, as the notice did not identify the date of the lease and incorrectly stated the name of one of the parties of that lease. Held that paragraph 6(3) rescued these defects, as the information provided was sufficient for the landlords to identify the correct leases.

(2) Failure to provide proper particulars of the tenancy sufficient to show that it was at all material times a tenancy at a low rent or treated as such. The tenant's response to that question simply cross referred to the information referred to above and stated the ground rent to be 35p per year. However there was no comparison between the 35p annual ground rent and the rateable value of the house as at 23 March 1965, which was necessary to show that the rent was less than two thirds of the rateable value on that date. Again the court accepted that this error was not sufficient to invalidate the notice in its entirety as it was a near certainty that 35p was less than two thirds of the relevant rateable value on that date.

(3) The tenant's failure to state the periods during the previous 10 years during which the tenant had occupied the house as her residence etc. The tenant's solicitor answered this question by the words "not applicable". In fact the question was very applicable as the information was necessary to establish that the tenants satisfied the "residence test" [since abolished for enfranchisement/extended lease claims by the CLRA]. The Court of Appeal held this omission to be fatal to the tenant's notice as (under the law as it was) that information was amongst the most important the tenant had to provide. Such information related to factual matters which might not be within the landlord's knowledge.

(4) Failure to supply additional information about the value of the house to demonstrate that it did not exceed the applicable financial limit for the exercise of the right. Again the answer given was "not applicable". No attempt had been made to provide the information. Again this was fatal to a tenant's notice.

- *Earl Cadogan* v *Strauss* [2004] 2 EGLR 69 — where in providing details of his current leases which commenced in April 1983, the tenant failed to provide details of earlier leases from 1972 which had been surrendered to enable the grant of the 1983 leases. Disclosure of the earlier leases was relevant as regards aggregation of the terms (see above) which in turn had a possible effect on eligibility. The Court of Appeal held that the error did not invalid the notice as the missing information was something which would have been known to the landlord if the landlord had investigated its own register of leases.

Failure to serve a correct notice at best means the tenant will have to start the process again with a new notice and at worst losing the right to enfranchise or extend a lease in its entirety. The worst case scenario was demonstrated in the *Speedwell* case as the existing 99 year leases (which had been granted at the beginning of the 20th century) were coming to the end of their contractual terms and the ground landlord had already served notices under schedule 10 to the Local Government and Housing Act 1989 (see Chapter 12) terminating their leases as at 1 July 1999 and proposing a new assured monthly periodic tenancies at a monthly rent of £225. The effect of those notices was to give the recipients only two months in which to notify the ground landlord of their desire to acquire the freehold under the 1967 Act. The fact that the notices subsequently served under the 1967 Act were found to be defective meant that the right to enfranchise or obtain an extended lease under the 1967 Act was now lost.

Landlord's counternotice

Paragraph 7 of the third schedule to the Act requires the landlord to reply to a tenant's notice in the statutory form stating whether or not the landlord admits the tenant's right to have the freehold or extended lease and if that right is disputed, to state the grounds on which it is disputed. At this stage the landlord must also disclose any intention on his part to apply for an order under sections 17 and 18 of the Act for an order entitling the landlord to resume possession of the house at the end of the natural expiry of the current lease on the grounds that the landlord requires the accommodation as his only or main residence or for an adult member of his family. Where these grounds for possession can be established, the leaseholder will be entitled to compensation.

Where a landlord's notice admits the tenant's right to acquire the freehold or extended lease, that admission will thereafter will be binding on the landlord unless the landlord can show that the tenant had misrepresented the facts of his claim or concealed material facts.

The form of landlord's counternotice is set out as Form 3 of the Leasehold Reform (Notices) Regulations 1997, which can be summarised as follows:

- an acknowledgement of receipt of the tenant's notice claiming the right to the freehold or extended lease as described
- stating whether the tenant's right is admitted or not admitted (and if not admitted on what grounds)

- stating whether the landlord intends to apply to the court for possession of the house under sections 17 or 18 upon the expiry of the current lease
- (if appropriate) reserving the landlord's right to object to the exclusion or inclusion to or from the tenants claim of other premises in which the landlord has an interest
- (in intermediate leasehold cases) stating that the notice is given by the landlord as "reversioner" for the purposes of schedule 1 to the Act.

Price for freehold

Section 9 of the Act deals with the calculation of the purchase price and associated costs of freehold enfranchisement together with the tenant's rights to withdraw from the transaction once the amount of the purchase price is known. As with collective freehold enfranchisement under the 1993 Act, it is recommended that a professional valuation be obtained based on the specific valuation criteria laid down by section 9.

The basic assumption on which purchase price is based is what a vendor of the landlord's interest might achieve on the open market (subject to the existing lease) but on the fictitious assumption that the tenant has no right to acquire the freehold or extend the lease (save in so far as the lease has already been extended). As with collective freehold enfranchisement, marriage value (see Chapter 9) will be split 50:50 between landlord and leaseholder and will be disregarded in its entirety if the unexpired term of the tenant's current tenancy exceeds 80 years. As well as paying the required purchase price for the enfranchisement, the leaseholder must reimburse the landlord's reasonable valuation and legal costs arising out of the exercise of the tenant's right. In addition the leaseholder must discharge any outstanding rent, service charges or other associated sums due to the landlord. If the amount of the purchase price is not agreed, section 21 (as amended) gives the LVT jurisdiction to determine the amount payable for the enfranchisement. However neither party may apply to an LVT to determine the price unless:

- the landlord has informed the tenant of the price he is asking or
- two months have elapsed without his doing so since the tenant gave notice of his desire to purchase the freehold under the 1967 Act.

Section 9(3) gives the applicant one month from determination of the price to withdraw from the transaction (on giving notice to the landlord) if he is unwilling or unable to acquire the house at the stated price — in which case:

- the notice claiming enfranchisement shall cease to have effect and the tenant shall be liable to compensate the landlord for any interference by the notice with the exercise of the landlord's power to dispose of or deal with his interest in the house or neighbouring property and
- prevents the tenant from serving a further claim for enfranchisement within the following 12 months.

Landlord's interest in mortgage

Section 12 of the Act states that on payment of the relevant price; the tenant will purchase the landlord's freehold title free from any mortgages affecting it. However, this will only occur if the tenant pays the purchase price (or a sufficient part of it) towards the discharge of the relevant mortgage. In practice the landlord's solicitor would normally provide the tenant's lawyer with a written undertaking to pay off any outstanding mortgage out of the money received from the tenant on completion of the transaction.

Extended leases

When a tenant of a house qualifies for an extended lease under the 1967 Act and gives the landlord written notice of his desire to have that extended lease, section 14 requires the landlord to grant to the tenant and the tenant to accept, in substitution of the existing lease, a new lease for the house and premises for a term expiring 50 years after the term date of the existing tenancy. That is to say that if the existing lease has 25 years to run, the replacement lease will be (25 plus 50) making 75 years in aggregate. However a landlord is not required to grant that replacement lease until the leaseholder pays to the landlord:

- reimbursement of a landlord's reasonable costs in verifying a leaseholder's right to an extended lease together with the costs of actually granting that lease and associated valuation costs in relation to the future rent payable for the extended lease and

- all sums due under the existing lease by way of rent, service charge or otherwise. Where the amount of any such sum is not fully ascertained, the leaseholder must offer reasonable security for the payment of such amount when it becomes quantified.

The landlord's requirement to grant the replacement lease is not affected in any way by the fact that the landlord's interest may be in mortgage and such replacement lease will be deemed to have been authorised by the lender. However, the landlord will be obliged to forward the tenant's signed counterpart replacement lease to the mortgage lender for safe-keeping with the mortgage deeds.

Similarly where it is the tenant's existing lease which is in mortgage, the replacement lease will automatically take subject to the existing mortgage and, having acquired the replacement lease, the leaseholder will be required to forward that document to the mortgage lender for safe keeping.

Section 15 deals with the rent and other terms of the replacement lease. The general rules are:

- that up to the contractual expiration date of the original lease (if it had not been extended/replaced) (the term date) the original ground rent will remain unchanged
- that as from the term date the new ground rent shall represent the notional letting value of the ground on which the house is built, making no allowance for the building itself
- that the new ground rent calculated on that basis will be reviewable, at the landlord's option, at 25 year intervals.

It will be seen that even during the extended period after the term date, the ground rent will be substantially less than open market value. Neither does the 1967 Act provide for the payment of any premium by the tenant for the extended lease.

The letting value itself for the extended period will not normally be calculated when the extended lease is granted (unless the previous lease was already about to expire) but is required by section 15(2)(c) to be determined no earlier than 12 months before the new rent is to take effect. The tenant will also bear the cost of obtaining a valuation for the new rent and any reviewed rent.

Section 15(3) makes clear that in addition to the new rent, the leaseholder will continue to be responsible for reimbursement of the landlord's costs in relation to services, repairs and insurance.

Section 16 makes clear that once a lease has been extended under the Act, there is no further right of extension. In this respect the provisions of the 1967 Act differ from lease extensions under the 1993 Act (flats and maisonettes), where a lease can be repeatedly extended.

Landlord's retained rights

Section 17 states that where a tenancy of a house has been extended under the 1967 Act, the landlord may at any time (but not earlier than 12 months before the original term/date) apply to the court for an order that he may resume possession of the property on the grounds that for the purposes of redevelopment he proposes to demolish or reconstruct the whole or a substantial part of it.

Where the landlord is able to substantiate such claim before the county court, the court shall declare that the landlord is entitled to repossess the house in return for a payment of compensation calculated in accordance with schedule 2 to the Act.

Where the existing lease has not been extended but the tenant would otherwise qualify for an extended lease and has given notice claiming such right, the landlord may apply to the court at that time for an order on the grounds that he proposes to demolish or reconstruct following the expiration of the existing lease. If the landlord's application to the court succeeds, the tenant's notice claiming enfranchisement will cease to have effect. Where the landlord succeeds in an application to the court, any notice given by the tenant relating to the acquisition of the freehold will cease to have effect if it is given after the date of the court order.

Section 18 of the Act gives landlords the right to defeat a tenant's claim to an extended lease in circumstances where on the expiration of the existing lease he reasonably requires possession of the property for his own occupation or an adult member of his family, and demonstrates this to the satisfaction of the court. But section 18 will only apply where a landlord had previously acquired his interest in the property before 19 February 1966. Where section 18 applies, the court will weigh up the interests of the landlord and the tenant and whether greater hardship will be caused by making the order than by refusing it. Where such an order is made the tenant will be entitled to compensation for the loss of the house.

Section 19 gave certain landlords (mainly the landed estates) power to seek ministerial approval to schemes under which they would retain powers of management over their estates to maintain adequate standards of appearance and amenity and regulate redevelopment. Therefore, even when the freehold of a former leasehold property is enfranchised under the 1967 Act, that property will remain subject to the section 19 scheme. Such schemes may empower the ground landlord to carry out work for the maintenance of a property, to regulate its redevelopment, use or appearance and to impose other obligations — and in exercising such rights; the ground landlord will have all the rights of a mortgagee.

When a Lease Comes to an End

What happens when a long residential lease comes to an end? Do the former leaseholders have to pack their bags and leave? In some cases "yes" — but in the majority of cases "no", but the former lease-holders will have to pay substantially more for their continued occupation of their property. In many cases the situation may have been avoided entirely by freehold enfranchisement — or postponed by a lease extension. But when the issue does arise, the rights and liabilities of landlord and former leaseholder will be governed by the provisions of schedule 10 of the Local Government and Housing Act 1989. Where schedule 10 allows a former leaseholder to remain in occupation, this will be at a full market rent. A landlord is only entitled to repossess at the end of the lease if he can establish certain statutory grounds and prove these to the court.

Continued occupation criteria

For a former leaseholder to qualify for continued occupation under schedule 10, the following criteria must be met.

- *The status of the previous lease*
  This must have been a long tenancy (meaning a lease for more than 21 years) at a low rent. A low rent is one which was less than two thirds of the rateable value of that property on 31 March 1990. In practice, at the time of writing, all expired leases for more than 21 years would have had to have commenced before 31 March 1990.

However this yardstick is no longer available for residential leases granted after that date as residential rateable values were abolished on 1 April 1990. Therefore the low rent test for any lease granted after 31 March 1990 is that the annual ground rent must not exceed £1000 for properties in Greater London or £250 elsewhere.

- *The assured tenancy test*
But for the fact that it was previously held on a long lease at a low rent, the occupancy must comply with the criteria for an assured tenancy within the meaning of Part I of the Housing Act 1988. An assured tenancy is the modern equivalent of the old Rent Act tenancy which gave a sitting tenant the right to occupy a residential property for life at a fair rent. That fair rent was always less than the equivalent open market rental value of the property. By contrast an assured tenancy gives the occupant lifetime security of tenure but at a full market rent, to be assessed in default of agreement by a rent assessment committee. An assured tenancy is also to be distinguished from a assured shorthold tenancy, which can normally be ended by a landlord at any time on as little as two months written notice. Only a private individual who occupies the property as his only or principal home can qualify for an assured tenancy. There are also other statutory exclusions relating to properties with a very high rateable value or very high rent, as well as a residential landlord exception and other specific exclusions.

Old lease continues until terminated

On the contractual expiration of a long residential lease, that lease will continue indefinitely under article 3 of schedule 10 until either the ground landlord or the former tenant serves statutory notice on the other party to bring it to an end.

Article 4 gives a ground landlord a general right to serve statutory notice terminating a long residential lease either on its contractual expiry (the term date) or at any time thereafter. That notice must be served not more than 12 nor less than six months before the date specified by the landlord for termination. However where, in relation to the landlord's notice, there is an issue pending before the court or rent assessment committee (see below) the date of termination cannot be earlier than three months from the date the issue is finally resolved. Note that in relation to schedule 10, an LVT is known by its old name as a rent assessment committee — although the two are one and the same.

A landlord's notice served under article 4 must be in the statutory format prescribed by the Long Residential Tenancies (Principal Forms) Regulations 1997 (as amended). To be legally effective, article 4(5) requires a landlord's notice to either:

- propose an assured monthly periodic tenancy and a market rent for that tenancy and stating that the terms of the tenancy shall in other respects be the same as those contained in the previous lease save such different terms as the landlord may propose or
- state that if the former leaseholder is not willing to give up possession on the specified termination date, the landlord proposes to apply to the court on stated statutory grounds for an order entitling the landlord to repossess the property. But that notice must also invite the former leaseholder (within two months) to notify the landlord in writing whether:
  - (a) in the case of a landlord's notice proposing an assured tenancy, the tenant wishes to remain in possession and
  - (b) in the case of a landlord's notice to resume possession, the tenant is willing to give up possession on the specified date.

Note also the right of a tenant under the Leasehold Reform Act 1967 or under the Leasehold Reform Housing and Urban Development Act 1993 to exercise the tenant's right to either enfranchise the freehold (whether individually or collectively) or seek an extended lease. Once a landlord's notice has been served bringing the existing lease to an end, the leaseholder has only two months in which to serve the appropriate notice exercising an individual right to enfranchise or extend the lease. In relation to collective enfranchisement, the time-limit is four months. If the tenant responds in this way with a claim for statutory enfranchisement or a lease extension, the landlord's notice will cease to have effect. In practice a former leaseholder who is able to obtain an extended lease under the 1967 Act may do so on better financial terms than by paying a full market rent under schedule 10.

Grounds for repossession

The grounds on which a landlord can defeat the right of a former lease-holder to remain in the property draw on those set out in schedule 2 to the Housing Act 1988, which provide grounds for repossession of any assured tenancy. The grounds are set out in article 5 of schedule 10.

- Ground 6 — that the landlord intends to demolish or reconstruct the whole or a substantial part of the dwellinghouse or carry out substantial works on any part of the property where:
  (a) intended work cannot reasonably be carried out without the tenant vacating because:
      (1) the tenant is not willing to agree to such a variation of the terms of the tenancy as would give such access and other facilities as would permit the intended work to be carried out or
      (2) the nature of the intended work is such that no such variation is practicable or
      (3) the tenant is not willing to accept an assured tenancy of such part only of the dwellinghouse as would leave in the possession of the landlord so much of it as is reasonable to enable the intended work to be carried out and, where appropriate, as required to give such access and other facilities over the reduced part as would permit the intended work to be carried out or
      (4) the nature of the intended work is such that a tenancy is not practicable and
  (b) either the landlord seeking possession acquired his interest in the dwelling house before the grant of the tenancy or that interest was in existence at the time of the grant and neither that landlord nor any other person has acquired that interest since that time for money or moneys worth.
- Ground 9 — That suitable alternative accommodation is available for the tenant or will be so available when the order for possession takes effect.
- Ground 10 — That some rent lawfully due from the tenant is unpaid on date on which possession proceedings are begun and were in arrear at the date of service of notice relating to those proceedings.
- Ground 11 — Whether or not rent is in arrear on the date proceedings for possession are begun, the tenant has persistently delayed paying rent which has lawfully become due.
- Ground 12 — Any obligation of the tenancy (other than one related to the payment of rent) has been broken or not performed.
- The condition of the dwelling house or any of the common parts has deteriorated owing to acts of waste by, or the neglect or default of, the tenant or any other person residing in the dwellinghouse and, in the case of any act of waste by, or the neglect or default of,

a person lodging with the tenant or a subtenant of his, the tenant has not taken such steps as he ought reasonably to have taken for the removal of the lodger or subtenant.

- Ground 14 — That the tenant or other resident or visitor has been guilty of conduct causing or likely to cause a nuisance or annoyance to someone residing, visiting or otherwise engaging in lawful activity in the locality — or has been convicted of using it for immoral or illegal purposes or has committed an arrestable offence in, or in the locality of, the dwellinghouse.

- Ground 14A — Which applies only when the landlord is a registered social landlord or charitable housing trust and the property is occupied by a couple living as husband and wife, in circumstances where one partner has vacated because of violence or threats of violence by the other towards that partner or another family member residing with that partner — and if the court is satisfied that the partner who has left is unlikely to return.

- Ground 15 — Which applies where the condition of any furniture provided for use under the tenancy has, in the court's opinion, deteriorated owing to ill-treatment by the tenant or another person residing in the dwellinghouse and, in the case of ill-treatment, by a lodger or subtenant, the tenant has not taken such steps as he ought reasonably to have taken for the removal of the lodger or subtenant.

Ground 6 is absolute in that once ground 6 has been proved, the court is obliged to make an order for possession. Grounds 9–15 (inclusive) are discretionary in that the court may refuse possession (whether on terms or otherwise) even if the statutory grounds have been proved.

Article 5 also specifies two additional grounds for possession, which, although absolute, are limited in effect. These are:

- article 5(1)(b) — On the grounds that for the purposes of redevelopment after termination of the tenancy, the landlord proposes to demolish or reconstruct the whole or a substantial part of the premises. Note however, that this statutory ground can only be relied on by a body to which section 28 of the Leasehold Reform Act 1967 applies.

- article 5(1)(c) — That the premises or part of them are reasonably required by the landlord for occupation as a residence for himself, an adult son or daughter or his spouse's parent. This ground is also limited in that it only applies to a landlord whose interest in the premises was purchased before 19 February 1966.

Interim rent

Several months are likely to elapse between service of a landlord's notice under article 4 bringing the former lease to an end and the date on which terms for the new assured tenancy are finally settled, provided that the landlord has not in the meantime substantiated grounds for possession of the premises. To prevent any unfairness to the landlord resulting from such delay, article 6 allows a landlord to serve notice on the tenant in the statutory form proposing an interim monthly rent to take effect from the specified date for termination of the existing lease and to continue until the interim arrangement is replaced by something more permanent. The landlord's notice proposing an interim monthly rent may be served with or after the landlord's notice under article 4 bringing the existing lease to an end.

On receipt of a landlord's notice under article 6 proposing an interim rent, the former tenant has two months in which to refer the issue to a rent assessment committee — failing which the landlord's proposed interim rent would take effect on the date stated.

When a referral is made to a rent assessment committee, the committee will determine an open market rent to take effect from the specified date of termination of the existing lease and on the assumption that the terms of the existing lease (other than as to the amount of the rent) will continue to apply to the new assured tenancy. Article 7 allows the landlord and the tenant to reach agreement between themselves as to the amount of the interim rent — in which case the agreed amount will have legal effect and the rent assessment committee will not become involved.

Tenant's right to terminate

There may be circumstances when a former leaseholder does not wish to remain in the occupation of the former leasehold premises under a market rent assured tenancy. In those circumstances article 8 allows the leaseholder to terminate the landlord and tenant relationship either on the contractual expiry of the existing lease or at any time thereafter, by giving to the ground landlord not less than one month's prior written notice. The fact that the landlord has already served notice under article 4 or that there has been an election by the tenant to retain possession, does not prevent the tenant from giving notice under article 8 terminating the tenancy earlier than the specified date of termination.

Terms of new tenancy

By article 9, the new assured tenancy will take effect on the statutory termination of the existing lease. The new tenancy will be a monthly tenancy and the rent for each month will be calculated in accordance with articles 10 to 12 of schedule 10. The schedule adopts a flip-flop procedure for determining the market rent, with any dispute to be resolved by the rent assessment committee.

As with the interim rent, the rent and terms proposed by the landlord in his article 4 notice will take legal effect unless the former leaseholder serves a counternotice on the landlord within the statutory two month period (article 10). If the tenant serves a counternotice within the two month period, that notice must set out the tenant's own proposals for the rent and other terms of the tenancy, in so far as these differ from those proposed by the landlord. Those terms will then take legal effect (in place of the landlord's previously quoted terms) unless within two months from receipt of the tenant's notice, the landlord refers the matter to a rent assessment committee for determination. Where such a referral has been made, the committee must address the issue in two stages (article 11).

First, the committee must decide whether there is any dispute as to the terms of the proposed tenancy other than those related to the rent, and ascertain what those disputed terms are. If there is a dispute as to the non-monetary terms of the tenancy, then the committee must resolve those issues first before deciding what the appropriate rent should be for a tenancy based on the resolved terms. Article 11 also sets out the basic principles on which the committee are to assess open market rental value. As with interim rent, article 12 allows the parties to agree matters between themselves in relation to terms of the proposed assured tenancy, in which case those agreed terms would take legal effect.

Landlord's application for possession

Where a landlord's notice to resume possession under article 4(5)(b) has been served on the tenant and either the tenant has responded within two months with an election to retain possession or the tenant would otherwise qualify as an assured tenant (see above), the landlord may apply to the court for an order establishing his right to possession of the premises at the termination of the existing lease, on proof of the relevant statutory grounds. But the court will not entertain such an application unless it is made within two months from receipt of the tenant's election to remain in possession (where applicable) or, where no election has been notified, within four months from the date of service of the landlord's notice under article 4.

If the landlord either fails to apply to the court within the relevant time-limit or if (having applied) the landlord's application fails, the landlord's notice to resume possession and anything done in pursuance of it, shall cease to have legal effect. Where that happens, article 15(4) gives the landlord a further month in which to serve notice on the tenant proposing an assured tenancy and stating the terms of that proposed tenancy.

Article 16 deals with the situation where, following on from a long residential lease at a low rent, there has been a subsequent lease of substantially the same premises also at a low rent. In those circumstances the later lease will be regarded for the purposes of schedule 10 as a long lease whatever its length.

Variation of Leases

A long residential lease can be varied in two ways:

- by a deed of variation agreed and executed between the parties (with the concurrence of any mortgagee either of the leasehold or the landlord's interest) or
- by application to an LVT by either party under Part IV of the Landlord and Tenant Act 1987 (but only where such variation is necessary to make satisfactory provision for anything associated with repair, maintenance, buildings insurance or service charges).

Note also that some lease variations are so fundamental as to amount, in law, as a deemed surrender by the leaseholder of the existing lease and its replacement by a new lease incorporating the revised terms. This might apply if it was proposed to extend or vary the period of the lease or substitute a different property.

Section 35 of the 1987 Act states that *any* party to a long lease of a flat may apply to an LVT for an order varying the lease in such manner as is specified in the application. The grounds on which any such application may be made are that the lease fails to make satisfactory provision with respect to one or more of the following matters, namely:

- the repair or maintenance of the flat or the building containing it or any land or building which is let to the tenant under the lease or in respect of which rights are conferred on the leaseholder
- the insurance of the building containing the flat or any other associated land or building

- the repair or maintenance of any installations (whether in the same building or not) which are reasonably necessary to ensure that occupiers of the flat enjoy a reasonable standard of accommodation
- t he provision or maintenance of any services reasonably necessary to ensure that the occupants enjoy a reasonable standard of accommodation (whether they are services connected with such installations or not and whether they are services provided for the benefit of those occupiers or services provided for the benefit of the occupiers of several flats)
- the recovery by one party to the lease from another of expenditure incurred or to be incurred for the benefit of that other party or several persons including that party
- the computation of a service charge
- such other matters as may be prescribed by regulations.

Factors determining what is a reasonable standard of accommodation may include the safety and security of the flat and its occupants and of the common parts of the building containing the flat and other factors relating to the condition of any common parts. Provisions relating to the computation of service charges will never be satisfactory if the aggregate of all the service charges paid by each leaseholder do not equate to exactly 100% of the landlord's expenditure. It will be seen that section 35 allows for an application to be made by a ground landlord as well as leaseholders.

As a variation of one lease may have a knock-on effect on other leases in the same building, Section 36 states that where an applicant applies for the variation of a lease, an applicant may also seek a variation of other long leases affected by the application. That application can only affect other leases which fail to make satisfactory provision in relation to the stated matters.

Section 37 enables a joint application to be made in relation to two or more leases for an order varying each of those leases. Those leases must be long leases of flats with the same landlord but need not be leases of flats in the same building or which are drafted in identical terms. The grounds on which an application may be made under section 37 are that the object to be achieved by the variation cannot be satisfactory achieved unless all the leases are varied to the same effect. Such application may be made by the landlord or any of leaseholders.

Where there are less than nine leases in a block, all, or all but one, of the parties must consent to it; or where there are more than eight leases, the proposed variation must not be opposed by more than 10% of the total number of parties concerned and at least 75% of that number must consent to it.

Section 38 states that if the grounds for variation are proved to the satisfaction of the LVT, the LVT may make an order varying the lease as specified in the application. However an LVT cannot order the variation of a lease if it appears to the LVT that the variation would be likely substantially to prejudice either the respondent to the application or anyone who is not a party to the application and that the effected person would not be awarded adequate compensation — or that for any other reason it would not be reasonable in the circumstances for the variation to be affected. An LVT cannot make any order which terminates any existing landlord's right to nominate an insurer or which requires the landlord to nominate a number of insurers from which the tenant would be entitled to select an insurer or which requires a leaseholder to effect insurance otherwise than with a specified insurer.

Instead of ordering a variation of a lease directly, Section 38 (8) allows an LVT to direct the parties to the lease to vary it in such manner as is specified. An LVT may also direct that a memorandum of any variation shall be endorsed on the lease or any other relevant documents. Where an LVT makes an order varying a lease, the LVT may, if it thinks fit, order any party to the lease to pay compensation to another party in respect of any loss or disadvantage that the LVT considers that other party is likely to suffer as a result of the variation.

Section 39 makes clear that any variation ordered by an LVT shall bind not only the current parties to the lease but also any other persons (including predecessors in title of those parties), whether or not they were parties to the proceedings to which the order was made or were with served prior notice of the application. Any variation effected by an order also binds any collateral guarantor of any leasehold obligation. Where someone who is not notified of the proceedings is bound by and suffers detriment as a result of that order, section 39(3) allows that person to bring an action for damages for breach of statutory duty against the person who failed to serve notice and may also apply to an LVT for the cancellation or modification of the order.

Disputes Resolution

There are two mainstream legal routes for resolving disputes between ground landlord and leaseholder. These are the county court and the LVT. Jurisdiction in landlord and tenant matters is split.

County courts will generally deal with the hard legal issues relating to the interpretation of documents; questions relating to eligibility; technical procedure; forfeiture and eviction; compensation claims; and the enforcement of rights. LVTs deal with the softer issues of valuation; qualitative assessment; reasonableness (in relation to service charges); and other issues requiring the expertise of a valuer or surveyor. With the exception of the small claims jurisdiction, proceedings before a county court remain adversarial and lawyer-driven. Proceedings before an LVT are more inquisitorial and members of the adjudication panel will draw upon their own professional experience to get to the core legal and valuation issues in the dispute. LVT procedures are mostly less formal and, in many cases, will include an inspection of the property by the adjudication panel.

As LVTs have evolved from the old rent assessment committees (who still determine fair rents in relation to older Rent Act tenancies — and market rents in relation to modern assured tenancies), one would assume that they would favoured by leaseholders. Whereas professional landlords may favour the more robust approach of the county court. Certainly an application to an LVT involves less risk for a leaseholder. Save where a party is shown to have behaved unreasonably, there is a general no-cost rule in LVT proceedings. It means that even if a leaseholder loses against a well represented landlord, the leaseholder will only be required to pay his own costs and those of the LVT but not those of the landlord. Neither does either party have to be represented by a lawyer, although it is desirable that each party has access to some form of professional representation, if not at the hearing itself, then in the preparation for the hearing.

Such representation could take the form of a surveyor or valuer, sufficiently familiar with LVT procedures to present an arguable case. This may have a double advantage if the surveyor can also provide relevant expert evidence. However where the value of a dispute is substantial, formal legal representation is recommended.

An LVT can only deal with those specific issues of valuation and qualitative assessment which have been expressively referred to it by statute. However the list is growing and, as will be seen from previous chapters, that jurisdiction now extends far beyond mere valuation of rents and premiums. If asked to do so, LVTs will now assess the reasonableness of proposed work; the adequacy of that work and whether it has been competently carried out; whether a leaseholder is in breach of any term of the lease; and whether a badly drawn lease needs to be varied. Issues relating to valuation are generally within the exclusive jurisdiction of an LVT. This means that a county court judge will not generally want to get involved in questions of valuation or adjudicate between competing expert opinions on such issues. He will instead refer such valuation issues across to an LVT for determination. Any other legal or factual issue for which statue has not given jurisdiction to an LVT will be within the exclusive jurisdiction of the county court. There are however many landlord and tenant issues in which the courts and LVTs share jurisdiction. In those cases, landlords and leaseholders will have some choice as to whether the issue is dealt with through the courts or through an LVT. This is of course subject to the court's power to transfer proceedings to an LVT, where it considers that venue to be more appropriate having regard to the overriding objective (see below). Issues in respect of which the courts and LVTs share jurisdiction in this way include:

- the reasonableness of service charges
- whether a leaseholder is in contravention of the terms of a lease.

At the end of this chapter we also look at other forms of alternative dispute resolution, which the parties to a dispute may adopt in place of the more formal proceedings before a court or LVT.

LVTs

The landlord and tenant issues for which LVTs have exclusive jurisdiction are set out below.

1. The capital valuation of a ground landlord's interest when a group of leaseholders collectively apply for freehold enfranchisement under the Leasehold Reform Housing and Urban Development Act 1993 (as amended).
2. The capital valuation of a landlord's freehold interest when the leaseholder of a house applies to enfranchise under the Leasehold Reform Act 1967.
3. The premium to be paid by an individual lessee for an extended lease under the 1993 Act.
4. The increased rental to be paid by a lessee for an extended lease under the 1967 Act.
5. Whether a manager should be appointed to take over a ground landlord's powers of management under Part II Landlord and Tenant Act 1987, following proof of landlord default.
6. The terms and conditions on which freehold enfranchisement or an extended lease should take place in the event of dispute.
7. The market rent and other terms of an assured tenancy to be granted by a former ground landlord to a former leaseholder on the expiration of a long residential lease under schedule 10 of the Local Government and Housing Act 1989. In exercising this jurisdiction the LVT is acting as a rent assessment committee.

LVT procedure is now codified in sections 173 to 175 of and schedule 12 to the CLRA as well as the Leasehold Valuation Tribunals (Procedure) (England) Regulations 2003 (as amended) and the Leasehold Valuation Tribunal (Fees) (England) Regulations 2003. A starting point is the 2003 Procedure Regulations (as amended).

Regulation 3 states that the particulars to be included with any application to an LVT must include:

*   the applicant's name and address
*   the respondent's name and address
*   the name and address of any landlord or tenant on the premises to which the application relates
*   the address of the relevant premises

- a statement that the application believes that the facts stated in the application are true (the statement of truth).

Schedule 2 to the regulations set out the additional documentation which is required for particular types of applications to an LVT.

- *Enfranchisement or extended lease* — include a copy of any notice served; details of the freeholder and an intermediate landlord; the name and address of any mortgagee; and a copy of the relevant lease.
- *Service charges, administration charges and estate charges* — include details of the secretary of any recognised tenants' association; where the application relates to the variation of any service charge provision, a draft of the proposed variation; a copy of the lease (or estate management scheme where relevant).
- *Estate management charges* — a copy of any estate management agreement or proposed scheme; a statement that the applicant is either an actual person, representative body or relevant authority; a copy of any relevant notice; a description of the area of the proposed scheme or modification or variation; and (where appropriate) a copy of any secretary of state consent.
- *Right to manage* — the name and address for service of the RTM company; the name and address of the freeholder and any intermediate leaseholder and manager; a copy of the memorandum and articles of association of the RTM company; a copy of the claim notice and counter notice (where relevant); a statement that relevant statutory requirements have been fulfilled; a statement that the participation notice has been served on all qualifying tenants; (where applicable) a statement describing the circumstances in which the landlord cannot be identified or traced; (where applicable) an estimate of the amount of accrued uncommitted service charges; (where appropriate) a copy of the relevant lease; (where appropriate) the date and circumstances in which the right to exercise the right to manage has ceased within the previous four years.
- *Appointment of manager* — (where appropriate) a copy of the notice served under section 22 of the 1987 Act or (where applicable) a copy of the management order.
- *Variation of leases* — the name and addresses of anyone served with formal notice together with a draft of the variation sought.
- *Determination of breach of covenant or condition* — include a

statement particularising the alleged breach of covenant or condition and enclose a copy of the relevant lease.

Regulation 4 requires the applicant to give notice of any application under Part 4 of the 1987 Act (variation of leases) to the other party to the lease and to anyone whom the applicant knows, or has reason to believe, is likely to be affected by the proposed variation. On receipt of such notice, the respondent must pass on notice to anyone not already notified whom, the respondent knows or has reason to believe is, likely to be affected by the proposed variation.

On receipt of any application (other than one made under part 4 of the 1987 Act — see above), Regulation 5 requires an LVT to copy the application and each of the accompanying documents to each person named in it as a respondent. On receiving an application relating to service charges, administration charges or estate charges, the LVT must also notify the application to the secretary of any recognised tenant's association and anyone else whom the tribunal considers is likely to be significantly affected by the application. Any person affected by the application may then seek to be joined as a party to the proceedings Regulation 5(6) also allows an LVT to advertise an application in two newspapers (at least one of which should be a freely distributed newspaper) circulated in the locality in which the premises are situated.

Regulation 6 states that anyone requesting to be joined as a party to proceedings must state whether they wish to be treated as a joint applicant or joint respondent to the application. The LVT then has a discretion either to grant or refuse that request. On reaching its decision on the request, the LVT must notify the third party and state the reasons for their decision and also send a copy of the notification to the applicant and the respondent.

Where a fee has not been paid as required by the Fees Regulations, regulation 7 prevents the LVT from proceeding further with the application until that fee is paid. Where a fee remains unpaid for more than one month from the date it becomes due, the application will be treated as withdrawn unless the LVT is satisfied that there are reasonable grounds for not doing so.

Where numerous applications have been made in respect of the same or substantially the same matter or include some matters which are the same or substantially the same, regulation 8 allows an LVT to determine only one of those applications ("the representative application") as representing all those applications and give notice of that proposal to all parties.

Notice given under regulation 8 shall: specify the common matters; specify the application which the LVT proposes to determine as the representative application; explain that the LVT's decision on the common matters will apply to all those applications; invite objections to the LVT's proposal to determine the representative application; and specify the address to which objections may be sent and the date (at least 21 days ahead) by which those objections must be received. In the absence of a timely objection, the LVT will determine the representative application and the result of that decision will apply to all the affected applications. Where a timely objection is received, that application will be dealt with separately.

Where a representative application has been determined and a subsequent similar application is received, regulation 9 requires the LVT to notify the later parties of the matters already decided in relation to the representative application and state that the LVT's decision on the common matters will also apply to the subsequent application unless any of the parties to the later application objects. If an objection is received, the LVT must consider the objection when determining the subsequent application but if the LVT dismisses the objection, it may record its representative decision as being the decision of the LVT with regard to subsequent application.

Regulation 11 enables the LVT to similarly dismiss any application which appears frivolous or vexatious or is otherwise an abuse of process. The respondent to such an application may also request that it be dismissed as being frivolous, vexatious or an abuse. But before dismissing any application under regulation 11 the LVT must first notify the applicant that it is minded to dismiss the application and the grounds on which it is intending to do so. The applicant then has 21 days to request to appear before and be heard by the LVT on the question of whether the application should be dismissed.

Regulation 12 allows an LVT, either on its own initiative or at the request of either party, to hold a pre-trial review in respect of the application. The LVT must give each party not less than 14 days notice (unless shorter notice is agreed) of the date, time and place of the pre-trial review. At that review the LVT shall:

(a) give any direction that appears necessary or desirable to secure the just, expeditious and economical disposal of the proceedings

(b) endeavour to secure that the parties make all such admissions and agreements as ought reasonably to be made in relation to the proceedings and

(c) record in any order made at the pre-trial review any such submission agreement or any refusal to make such admission or agreement.

Pre-trial reviews may be exercised by a single member of the LVT.

Regulation 13 (as amended) allows an LVT to determine an application without a hearing where:

(a) it has given both parties not less than 28 days written notice of its intention to proceed without a hearing and
(b) neither party has made a request to the LVT to be heard.

Once a decision has been made to deal with the application without an oral hearing, the LVT will:

(a) notify the parties that the application is to be determined without an oral hearing
(b) invite written representations on the application
(c) set time-limits for sending any written representations to the tribunal and
(d) set out how the tribunal intends to determine the matter without an oral hearing.

However, at any time before the application is determined, either party may request a hearing or the LVT may decide to proceed by way of a hearing.

Regulation 14 deals with the hearing itself. The LVT must first give each party not less than 21 days prior notice of the scheduled date, time and place of the hearing. In exceptional cases the LVT may give less than 21 days notice, but any such notice must be given as soon as possible before the appointed date and specify what the exceptional circumstances are. An LVT may also arrange that an application be heard together with one or more other applications.

Regulation 14(6) states that the hearing shall be in public unless, in the particular circumstances of the case, it decides that hearing or part of hearing shall be heard in private. Regulation 14(7) entitles an LVT to decide what procedure to adopt and anyone appearing before an LVT may do so either in person or by representative authorised by him, whether or not that representative is a barrister or a solicitor. Anyone appearing before an LVT may also give evidence on his own behalf, call witnesses and cross-examine any witnesses called by another party. If one of the parties does not appear at the hearing, the LVT may proceed in their absence if satisfied that notice has been given in accordance with the regulations.

Regulation 15 entitles an LVT to postpone or adjourn a hearing or pre-trial review either on its own initiative or at the request of a parties. But where a postponement or adjournment has been requested, the LVT must not postpone or adjourn the hearing except where reasonable to do so having regard to: the grounds for the request; the time at which the request is made; and the convenience of the parties. The LVT must give reasonable notice of any postponed or adjourned hearing to each party.

Regulation 16 requires an LVT to take all reasonable steps to ensure that each party is given copies of all documents relevant to the proceedings. If at a hearing a party has not previously received a relevant document or copy or sufficient extract or particulars, then unless that person consents to the continuation of the hearing or the LVT considers that the party has a sufficient opportunity to deal with the matters to which the documents relate without an adjournment of the hearing, an LVT must adjourn the hearing for a sufficient period to enable that person to deal with those matters.

Regulation 17 allows an LVT (by appointment) to inspect the house, premises, or area which is the subject of the application or any comparable house, or premises or area to which its attention is directed. Each party must also be given an opportunity to attend the inspection. Where an inspection is to be made before a hearing, the LVT must give at least 14 days prior notice to each party stating the intended time, date and place. Where an inspection is made after the close of a hearing, an LVT may reopen the hearing in light of any matter arising during the inspection.

Regulation 18 deals with the issue of decisions by an LVT. Where a hearing is held the decision *may* be given orally at the end of the hearing. However that decision must always be recorded in a document as soon as possible after the decision has been made. The reasons for a decision must also be stated either in the decision notification itself or in a separate document issued as soon as possible after the decision has been recorded. A document recording an LVT decision must be signed and dated by a member of the LVT authorised to sign-off such decisions. Regulation 19 makes clear that any decision of an LVT may, with the permission of the county court, be enforced in the same way as orders of the court.

Appeals against LVT decisions

Section 175 allows appeals from an LVT decision to go to the Lands Tribunal — but such appeal can only be made with either with the permission of the LVT who made the decision or the Lands Tribunal itself and such appeal must be made within the times specified by rules under section 3(6) Lands Tribunal Act 1949.

Regulation 20 of the Procedure Regulations states that where a party applies to the LVT for permission to appeal to the Lands Tribunal, the application must be made to the LVT within 21 days starting with the date on which the document which records the reason for the decision was sent to the aggrieved party; and a copy of that application will then be served by the LVT on every other party.

Fees and costs relating to LVT proceedings

As stated, the general rule is that each side bears their own costs in LVT proceedings. However schedule 12 paragraph 10 allows an LVT to require a party to pay an opponent's cost in circumstances where:

* an application was dismissed because of an applicant's failure to pay the relevant LVT fee or
* the relevant party has, in the LVT's opinion, acted frivolously, vexatiously, abusively, disruptively or otherwise unreasonably in connection with the proceedings. Even when costs are awarded against a party who has behaved badly, regulation 10(3) caps those costs at £500.

The Leasehold Valuation Tribunals (Fees) (England) Regulations 2003 set out the fee structure for LVT applications. There is an initial fee payable by the applicant when lodging the application. A second fixed fee of £150 (again payable by the applicant) falls due when the application is set down for hearing. There are different fees for different types of application and these may vary according to numbers of properties involved.

For service or administration charge disputes, there is a rising scale of fees from £50 for the smallest disputes to £350 for claimed service charges, insurance premiums or administration charges exceeding £15,000.

Fees for the appointment of managers or variation of leases or determinations as to the suitability of insurers start at £150 (for less than six dwellings) rising to £350 for more than 10 dwellings.

Anyone receiving means tested state benefits is exempted from having to pay any LVT fee by regulation 8.

Regulation 9 allows an LVT to require any other party to the proceedings to reimburse the applicant for the whole or part of any of the fees paid by him. This is entirely separate from an LVT's power to award costs (see above).

An LVT cannot require a party receiving means tested benefits to reimburse another party's fees.

County court

County court procedures are governed by the Civil Procedure Rules and associated Practice Directions, which are accessible from the Court Service website *www.hmcourts-service.gov.uk*

Where a party is claiming damages or other relief, such as an injunction, declaration or specific performance of a contract, the court will expect the claimant to have previously written to the other party explaining their claim and seeking their formal response before proceedings are issued. There are also pre-action protocols for certain types of claim, which can be accessed from the court website.

But not all legal proceeding involve claims for damages or other legal redress. In many landlord and tenant situations it is essential to institute proceedings within statutory deadlines simply to preserve a leaseholder's rights. Examples are the statutory procedures relating to freehold enfranchisement or lease extension which (in the absence of agreement) require applications to made either to the court, or where appropriate, to an LVT.

Instituting court proceedings

Most court proceedings are commenced by the issue of a claim form, setting out the grounds of the claim and what redress the claimant is seeking. For certain non-monetary claims, the claim form must be issued under part 8 of the rules.

Once a claim form has been issued and served on the other party, that party has 14 days first to acknowledge receipt of the claim form and to state whether or not the claim is to be defended — and (if a formally acknowledgement is made) a further 14 days to issue a formal response to the claim. If the defendant (or respondent) either fails to acknowledge the claim form, or subsequently fails to issue a detailed response, the general rule is that the claimant can apply immediately for judgment to be entered against the defendant/ respondent for the amount of the claim as stated. Where that claim has already been quantified in monetary terms, judgment is entered automatically at the claimant's request. Where the amount of damages to which the claimant is entitled, or other legal redress, has yet to be assessed by the court, judgment will be entered in principle, subject to assessment of the appropriate legal redress. Where the lodgement of court proceedings is merely part of a wider statutory process, as in the right to manage, enfranchise or extend a lease, special procedures and deadlines will apply, and non-compliance with any can have serious consequences for the party in default.

After the issue of a claim form and the receipt of a detailed response (and subject to receipt of any other information about the claim or defence which is required by the court to identify the particular issues in dispute), the court will ask each party to complete an allocation questionnaire, seeking information about the estimated value of the claim (and any counter claim), numbers of witnesses to be called, time estimate and other information indicating the complexity of the case. On receipt of the completed questionnaire, a district judge will allocate the proceedings to one of three tracks, namely: the small claims track; the fast track; or the multi track.

The tracking of a case has significance for the parties as it (and any associated court directions) will state the procedure which the court adopts to deal with the case, which may in turn affect the likely costs of those proceedings.

Small claims track

The county court small claims procedure can provide the quickest, cheapest and simplest way to resolve a small monetary dispute.

The procedure is informal, self-explanatory and suited to someone presenting their own case. Although legal representation is permitted, long speeches and complex legal argument are not welcomed. Although technically open to the public, a small claims hearing will usually take place around a conference table in the district judge's office. The pace is quick, with most small claims hearings scheduled to last no more than two hours, and the district judge will lead from the start to get to the main issues in the dispute. Document disclosure is also kept to a minimum, with each party only being required to disclose those documents which they regard as essential to the presentation of their case.

Like the LVT, small claims proceedings are comparatively risk free for an unrepresented claimant or defendant as again there is a general no-costs rule, unless one of the parties is shown to have behaved unreasonably in the proceedings. The small claims track is also quick in that once the case has been allocated to that track, notice of a hearing date will be issued almost immediately — with the hearing itself taking place two to three months hence.

As well as awarding monetary compensation, district judges exercising their small claims jurisdiction can also issue injunctions or orders for specific performance — but not emergency injunctions or other interim remedy. Although the small claims jurisdiction generally covers monetary claims up to £5,000, its jurisdiction in personal injuries and housing disrepair claims is more limited:

- the small claims jurisdiction will not apply where a claim for personal injuries exceeds £1,000
- the small claims jurisdiction will not apply in relation to a claim by a residential tenant against a landlord for repairs or other work to the premises where the estimated cost of the repairs of other work exceeds £1,000 or the financial value of any other claim for damages exceeds that amount.

Fast track

The fast track generally covers claims between £5,000 and £15,000 where the trial is not estimated to last longer than one day. Normal court rules will apply save that there is a limitation on the amount of costs which the losing party is required to pay the winner. This is to encourage each party to pursue their case economically. The fast track will also deal with personal injury and housing disrepair claims which escape the small claims track because they exceed £1,000. Note also that although (in value terms) a claim may be within the small claims jurisdiction, the court may move it into the fast track (or even the multi track) if there are particularly complex issues which need to be resolved. This is a risk which any small claims litigant has to face.

Multi track

This track covers all civil claims, of whatever nature, which do not fall within the small claims track or fast track.

Transfer of proceedings

Paragraph 3 of schedule 12 of the CLRA states:

(1)   where in any proceedings before a court there falls for determination a question falling within the jurisdiction of a Leasehold Valuation Tribunal, the court:

   (a)   may by order transfer to a Leasehold Valuation Tribunal so much of the proceedings as relate to the determination of that question, and

   (b)   may then dispose of all or any remaining proceedings, or adjourn the disposal of all or any remaining proceedings pending the determination of that question by the Leasehold Valuation Tribunal, as it thinks fit.

(2)   When the Leasehold Valuation Tribunal has determined the question, the court may give effect to the determination in an order of the court.

It clear from schedule 12(3) that a court has a discretion whether to transfer (or not transfer) relevant questions to an LVT, even when the LVT has parallel jurisdiction in the matter. In deciding whether or not to transfer proceedings to an LVT, the court will have regard to the overriding objective defined in part 1 of the Civil Procedure Rules which states:

1.1   These Rules are a new procedural code with the overriding objective of enabling the court to deal with cases justly.

   (2)   Dealing with a case justly includes, so far as practicable –

      (a)   ensuring that the parties are on an equal footing;

      (b)   saving expense;

      (c)   dealing with the case in ways which are proportionate –

         (i)   to the amount of money involved;

         (ii)   to the importance of the case;

         (iii) to the complexity of the issue; and

         (iv)   to the financial position of each party;

      (d)   ensuring that is it dealt with expeditiously and fairly; and

      (e)   allotting to it an appropriate share of the courts resources, while

      (f)   taking into account the need to allot resources to other cases.

1.2 The court must seek to give effect to the overriding objective when it —
    (a)   exercises any power given to it by the Rules; or
    (b)   interprets any rules.
1.3 The parties are required to help the court to further to the overriding objective.

Paragraph 1.4 requires the court to further the overriding objective by actively managing cases. Such active case management is expressed to include (among other things): encouraging the parties to co-operate with each other in the conduct of the proceedings; helping the parties to settle the whole or part of the case; and encouraging the parties to use an alternative dispute resolution procedure if the court considers that appropriate and facilitating the use of such procedure. Use of alternative dispute resolution (or ADR) is referred to at the end of this chapter.

The issue of whether a case, which had been started in the county court, should be transferred to an LVT, was considered by the Court of Appeal in *Aylesbond Estates Ltd* v *MacMillan* (LTL24/11/98).

The ground landlord had sought forfeiture of a 99 year lease of Flat 9, Sinclair Road, London W14, on the grounds that rent and service charges had not been paid. The landlord issued proceedings in the West London County Court on 15 September 1994. Almost four years later those proceedings were still pending and no trial date was fixed until July 1998. The trial was then adjourned because one of the joint leaseholders was ill and the other had a prearranged appointment elsewhere. Following a subsequent abortive hearing, the trial was fixed for hearing on 5 and 6 August 1998.

In the meantime the leaseholders applied for the transfer of the proceedings to Edmington County Court in the expectation that the judge of that court would transfer the proceedings on to an LVT, as there was a dispute over service charges. One of the reasons why the leaseholders wanted to transfer to an LVT was that the costs of the current proceedings were estimated to be in the order of £20,000–£30,000 and even if those costs could not be recovered directly from the leaseholders, they could be recovered by way of additional service charge against other tenants. The leaseholders believed that such a result would be best avoided by the case being transferred to the LVT, where the leaseholders believed that the costs would be considerably less than if the case were to proceed to trial in the county court. The trial judge refused to transfer the proceedings and the leaseholders appealed to the Court of Appeal on this specific issue. The Court of

Appeal said that the judge was right to refuse transfer to an LVT.
Tuckey LJ said;

> they [the LVT] cannot determine disputed questions of forfeiture which
> arise in this case since it is the landlord's case that the lease was forfeited
> when there was a failure to make payment; nor can the tribunal deal with
> resulting questions of relief from forfeiture which may arise in this case.
> Disputes about ground rent are likewise not the subject of the jurisdiction of
> the LVT, and the counterclaim of the [leaseholders], in so far as it relates to
> sound insulation and damages for the breach of the covenant of quiet
> enjoyment and nuisance, is outside the jurisdiction of the LVT. I note in
> passing that, recognising the difficulty, the [leaseholders] indicated to us
> that they would be prepared to forgo their claim for damages in respect of
> the lack of sound insulation in the flat. But there are other issues which arise
> to which I have referred which cannot be decided by the LVT. So this is a
> case which the court *may* refer the service charge dispute to the LVT but if
> so it would be in the knowledge that this would by no means dispose of the
> entire dispute between the landlords and the appellant.

Later in his judgment, Tuckey LJ added

> it seems to me that there are no good costs reasons why this can be sent to
> the LVT. I bear in mind of course that this is an expert tribunal, consisting,
> as we are told, of a lawyer and two experts in this field, and of course that
> might mean, following an inspection the tribunal, could reach a decision on
> the service charge dispute more quickly and easily, and therefore less
> expensively, by a judge that does not have this experience. But against that
> must be said the fact there are not, as far as this dispute is concerned,
> currently any proceedings before the LVT. If this court were to refer the
> dispute to the LVT, I have no doubt that they would wish to look at the
> position to decide what directions to make to aid their determination and so
> on. That would be a somewhat lengthy and somewhat costly exercise, so
> that any advantage to be gained by the expertise of the tribunal in terms of
> costs saving would be off-set by the fact that process, so far as this part of
> the dispute was concerned, would have to start again. But that is not the
> determining factor, in my judgment. It seems to me, that so far as the county
> court is concerned, a large part of the costs have already been incurred.
> Twice the landlords have been at the court door ready to proceed. Witness
> statements have been exchanged. Discovery has been completed. Much of
> the cost of those proceedings has therefore already been incurred. If one is
> looking to see where these parties could best expect to have a quick
> resolution of their dispute, I have no doubt whatsoever that the answer to
> that question must be in the county court, and that, it seems to me, is the
> determining factor, there being no obvious costs advantage in going to the

LVT. The advantage of a speedy resolution in the county court as compared with the LVT plus the fact, that were this case be referred to the LVT, that would by no means be the end of the matter, whereas in the county court all of the issues between the parties, which I have identified, can be sorted out by one trial, which I hope will be very soon, are the decisive factors in my judgment.

Alternative dispute resolution

As stated, the courts' overriding objective itself encourages alternative dispute resolution (ADR) wherever possible. In practice ADR generally means one of two things:

- arbitration or
- mediation.

The procedures and structure of an arbitration may resemble that of a court in that evidence is presented (whether in writing or orally) and the arbitrator makes a binding decision based on that evidence. The decision of the arbitrator is generally final unless there are either procedural irregularities or the decision is wrong in law. The entire procedure is governed by the Arbitration Act 1996 and associated regulations.

The arbitration procedure is voluntary — in that no party can be forced to go to arbitration in relation to service charges even if the lease so provides. However in the context of a residential lease, the courts will recognise a post-dispute arbitration agreement — where the parties have chosen the arbitration route as a means of resolving a dispute which has already arisen. The arbitrator will be appointed either by agreement between the parties or selected by the President for the time being of the Law Society or Royal Institution of Chartered Surveyors (or some other eminent institution). The procedure can also be expensive as the arbitrator, being a professional adjudicator with special expertise, will also require payment.

Mediation differs from arbitration in that the mediator does not impose any solution on the parties. Instead a third person (acting as facilitator or go-between) helps the parties to reach their own amicable solution to the dispute. Mediation can assist constructive negotiation and settlement in a dispute, even in situations where personal relations between landlord and leaseholder have broken down so badly that they will hardly talk to each other. The aim of mediation is first to address the personalities before moving on to address the actual issues in dispute. Mediation is particularly useful in a situation (like landlord and tenant) where the parties are locked into a long term legal relationship and will have to deal with each other in the future. Unlike arbitration, mediation is only final if the parties are able to negotiate an amicable solution to their dispute and produce a written agreement recording those terms. If no solution is reached, either party is free to take their dispute to the court or LVT (where applicable). The suggested starting point for any mediation is a telephone call to the Centre for Effective Dispute Resolution (CEDR) whose details can be found on the Internet. One of their advisors will then explain the procedure and quote a price.

Running a Leasehold Management Company

Unless among their number, the leaseholders include a lawyer or accountant with special expertise in such matters, it must be assumed that the services of a solicitor or accountant will have been engaged in the setting up of any leaseholder's management company. That company may have been set up as part of a wider statutory process involving the collective right to enfranchise or take over the management of a block of flats. Once the company has been set up, how much more professional expertise the leaseholder/participants require to run the company depends to a large extent on their own expertise, time, energy — and most of all, their commitment to the company. The more leaseholders can do themselves, the cheaper the overall running costs of the company. This final chapter provides an over-view of what running a management company entails.

Preliminary meeting

It is recommended that before any formal step is taken towards formation of a company, the leaseholders meet to decide:

- a name for the company
- the structure of the company (in so far as the leaseholders have any choice over this)
- the registered office of the company
- an assessment as to the likely cost of setting up the company and the future cost of continuing to run it over the longer term

- what additional professional help is required in the setting up and future running of the company and who should be appointed to provide such professional help.

The general rule is that anyone can call their company anything they like. But a company name must not be similar to that of another registered company. A check of the Companies House website will reveal whether there are other companies with similar names.

Some company names are regarded as sensitive and will be refused by Companies House. These are names which may suggest some inappropriate official connection, such as the use of "Royal". For example, one management company was refused the word "Kings", even though the name of the road in which the block was situated was Kings Avenue. Remember also that right to manage and right to enfranchise companies must include the words RTE or RTM in their company names.

The registered office of a company may be the address of one of the leaseholders involved (usually that of the company secretary) or a professional address such as a solicitor or accountant. It is to the registered office with whom Companies House will always correspond and to whom any other official documents sent to the company must be addressed.

The constitution of the company

The memorandum of association and the articles of association jointly comprise the constitution of any registered company. It is those documents which set out the powers of the company, the liability of individual members, restrictions on membership, appointment of directors and company secretary, and how decisions are to be made. It is therefore important to ensure that all meetings are convened and decisions made strictly in accordance with the provisions of the constitution, otherwise they will not be recognised in law as being decisions of the company. But the memorandum and articles of the company do not sit in isolation as, in the case of a leaseholders' management company, they will have to be read together with the terms of a third document, namely the lease of each individual flat.

The roles and responsibilities of directors

The directors are elected from among the membership of the company and are responsible for its day to day business. Their decisions will be regarded as decisions on behalf of the whole company. One of those directors will be the chairman (chairperson or simply chair) of the company. That person will preside at meetings of the directors or wider membership and also be the company's spokesperson. That person will also work closely with the company secretary in the setting of agendas and arranging of meetings.

The concept of limited liability means that directors and officers of a company should not be personally liable for its debts, even if things go wrong. But there are circumstances in which directors and company secretaries can be held personally to account if they act negligently, in bad faith or outside the powers delegated to them by the articles of association. Insurance for company directors and officers can be obtained against such liability. But there is a restricted market for such insurance and the costs of obtaining it may be disproportionate to actual risk involved. The best way for directors and officers to minimise any personal risk to themselves is to know when to take professional advice, to take that advice and then to act in accordance with it.

Where the membership of a management company is small, every leaseholder/participant in the company may double up as a director — in which case, director's meetings will be almost indistinguishable from meetings of the wider membership. For large companies, directors will be elected at the annual general meeting from among the wider membership.

Role of company secretary

The company secretary is the chief administrative officer of the company. It is the company secretary who "behind the scenes" will provide the backbone of the company and carry the biggest and most time-consuming workload. The company secretary can be a professional person such as a lawyer or accountant — or alternatively can be one of the leaseholder/participants in the company. Either way the company secretary must be fully committed to the company, have the time to deal with its administrative affairs, be a good administrator and have some bookkeeping experience. The roles of a company secretary include:

- keeping the statutory books of the company and keeping those books up to date. The statutory books are: the Register of Members; the Register of Directors and Secretaries; the Register of Directors' Interests; and the Register of Charges. Added to this must be a Minute Book, providing a formal signed record of all decisions made by the company
- in liaison with the company chairman, to arrange meetings and set the agendas for those meetings. The Articles of Association will set out the time-limits for sending those agendas to individual members or directors of the company
- keeping accounts
- filing tax returns
- compliance with Companies House formalities including the filing of accounts; filing annual returns; and providing official notification of any changes affecting the company
- writing up the minutes
- dealing with day to day correspondence.

In addition, for a leaseholders' management company, a company secretary will also have the following responsibilities:

- invoicing for half-yearly ground rents in the statutory format
- complying with the procedures set out in Chapter 5 in relation to major service charge items.

Procedure at meetings

Every leaseholders' management company should hold an annual general meeting to which the whole of its membership must be invited. The primary purpose of the AGM is to deal with formal matters such as the appointment of directors and a company secretary for the forthcoming year; appointment of any solicitors, accountants or managing agents; approval of accounts — before moving on to any specific items on which decisions need to be made. The AGM may also be used to specify the dates and times of future meetings during the course of the year. Other meetings can then be convened in accordance with the articles of the company. The general procedure for any company meeting is set out below.

1. That the pre-meeting formalities should have been complied with,

including the sending out of agendas to each member or director specifying the decisions, which it is proposed should be made.

2. Ensuring that a quorum is present at the start of a meeting. A quorum is the minimum number of members required to be present before any decisions can be made which are binding on the company. For example, if a company has eight members and three are specified as the quorum, the meeting cannot start until at least three members of the company are present. If a quorum is not present at a start of the meeting it is usual to wait for up to half an hour to see if other members attend, before starting the meeting. If however, it becomes apparent that there will not be a quorum of members, there may be no alternative but to postpone the meeting to a date and time when more members can attend.

3. For the chairman to open the meeting and for an attendance list to circulate. Non-attendees may have offered their apologies for absence.

4. The first formal business will be the approval of the minutes of the previous meeting. A draft of those minutes should have been appended to the agenda sent to each member or director. Once those minutes are approved by resolution of the company and signed by the chair of the meeting as being an accurate record of decisions made at the previous meeting, those minutes will become official evidence of matters already resolved.

5. The next item may be matters arising from previous minutes — which gives the membership an opportunity to catch up with what was agreed and what has happened in relation to those decisions.

6. The formalities having been dealt with, the meeting then moves on to make specific decisions in relation to each item on the agenda. Where possible, unanimous agreement should be reached by everyone present. Where unanimous agreement is not possible a vote may have to be taken. That vote would normally be on a show of hands, with the chairman having a casting vote in the event of a tie. But note the special voting arrangements which exist for RTM companies, whereby some members may have more votes than other members, according to a statutory formula. Where a vote is taken, a detailed record must be made of those voting "for" or "against" the motion. If there is a majority in favour of the motion, it becomes a resolution of the company. If the vote is lost, discussion on the item continues until an alternative motion is put forward.

7. The final item on the agenda is normally '"any other business". This is not an opportunity for the company to make major decisions which should have been included on the agenda. Instead it provides an opportunity for anyone present to raise any urgent items or express any view as to what issues the company may have to address in the future.

8. Following the close of the meeting, the company secretary will write up the minutes and send them, in the first instance, to the chairman, for comment. Those draft minutes will then be attached to the next agenda for approval at that meeting.

Compliance

Every year Companies House will send out an annual return for completion and return, to update Companies House records in relation to the company. In addition, annual accounts must be provided to Companies House. This accounting information will become a matter of public record to which anyone has access. Most leaseholders management companies rank as "tiny", which means that the accounts can be submitted in an "abbreviated" form, comprising little more than a balance sheet. Neither will those accounts have to be audited. However those accounts must be approved by the membership of the company and must be presented to Companies House in the statutory format. Where company secretaries do not have sufficient bookkeeping expertise themselves to deal with this, the services of an external accountant will be required. There will also be Inland Revenue Tax Returns to be completed in respect of the company.

Where annual returns or accounts are not filed in accordance with statutory requirements, each of the directors and the company secretary will be exposed to the risk of prosecution and financial penalties. Furthermore there are automatic penalties imposed on the company itself when there is a delay in providing such documentation.

Index